PENGUIN BOOKS

RANCH OF DREAMS

A national television program recently called Cleveland Amory "the premier protector of animals on this planet." Prior to his activities with animals, however, Amory was well known for his bestselling books on social history, *The Proper Bostonians, The Last Resorts,* and *Who Killed Society?* A longtime columnist for *The Saturday Review,* chief critic for *TV Guide,* and a contributor to *Parade Magazine,* Amory is now best known for his three cat books, *The Cat Who Came for Christmas, The Cat and the Curmudgeon,* and *The Best Cat Ever,* all of which were national bestsellers. The founder and president of the Fund for Animals, Amory now spends most of his time at the Fund's headquarters at 200 West 57th Street in New York City, but he is also often found at the Fund's Black Beauty Ranch in Murchison, Texas. There he is regularly visited by his favorite burro, Friendly, now over twenty years old, who, whenever he appears, clumps faithfully up the steps of the porch for tea, tidbits, and gossip.

RANCH

OF

DREAMS

A LIFELONG PROTECTOR OF ANIMALS SHARES THE STORY OF HIS EXTRAORDINARY SANCTUARY

Cleveland Amory

PENGUIN BOOKS

PENGUIN BOOKS
Published by the Penguin Group
Penguin Putnam Inc., 375 Hudson Street,
New York, New York 10014, U.S.A.
Penguin Books Ltd, 27 Wrights Lane,
London W8 5TZ, England
Penguin Books Australia Ltd, Ringwood,
Victoria, Australia
Penguin Books Canada Ltd, 10 Alcorn Avenue,
Toronto, Ontario, Canada M4V 3B2
Penguin Books (N.Z.) Ltd, 182–190 Wairau Road,
Auckland 10, New Zealand
Penguin India, 210 Chiranjiv Tower, 43 Nehru Place,
New Delhi 11009, India

Penguin Books Ltd, Registered Offices:
Harmondsworth, Middlesex, England

First published in the United States of America by Viking Penguin,
a member of Penguin Putnam Inc. 1997
Published in Penguin Books 1998

1 3 5 7 9 10 8 6 4 2

THE LIBRARY OF CONGRESS HAS CATALOGUED THE HARDCOVER AS FOLLOWS:
Amory, Cleveland.
Ranch of Dreams / Cleveland Amory.
p. cm.
ISBN 0-670-87762-X (hc.)
ISBN 0 14 02.6975 4 (pbk.)
1. Amory, Cleveland—Homes and haunts—Texas. 2. Authors, American—20th century—Biography. 3. Animal rights activists—United States—Biography.
4. Black Beauty Ranch (Tex.) 5. Animal sanctuaries—Texas.
6. Animals—Treatment. I. Title.
PS3551.M58Z47 1997
818′.5403—dc21 97-29731

Printed in the United States of America
Set in Centaur
Designed by Liney Li

This book

is dedicated to all those

whose hearts burn,

as does that of the author,

with a deep and abiding

hatred of cruelty

to animals.

ACKNOWLEDGMENTS

The author wishes to acknowledge the invaluable assistance of his longtime assistant, Marian Probst, whose incredible memory, hard as it is to bear at certain times, was, in the matter of this book, an essential ingredient. Second, the author wishes to thank Carolyn Carlson, his editor at Viking Penguin, and also her assistant, Michael Driscoll, both of whom were in the author's opinion like so many people nowadays entirely too young for the job but nonetheless gave generously of their time and effort. Third, the author wishes to acknowledge the extraordinary efforts of Sean O'Gara, whose advice in the preparation of this manuscript was immeasurable even if it came at the expense of having to listen to what for a Boston Red Sox fan is intolerable—pro-Yankee propaganda. Fourth, the author wishes especially to thank his longtime friend, Walter Anderson, the editor of Parade, *whose support for the author's animal books has been unfailing and unstinting, albeit the author cannot get out of his mind the feeling that Mr. Anderson would like them shorter. Fifth, the author wishes to thank, if you can believe it, not one but two lawyers—Edward Walsh and George Sheanshang, both of whom, in their lawyerly way, were at certain times and for short periods vitally unlawyerly. Finally, the author would be derelict in his duty if he did not mention the constant vigilance of one who, day in and day out and night in and night out, walked back and forth over the manuscript, at the same time keeping a weather eye out for both the book's progress and its dreadful impending deadline—his beloved office cat, Polar Star.*

CONTENTS

RANCH OF DREAMS

Chapter One

OF A MAIDEN AUNT, A GREAT-UNCLE, AND A BOOK BY A HORSE

ANIMALS HAVE BEEN SPECIAL TO ME FOR AS long as I can remember. In fact before I can remember we had a Cocker Spaniel named Duke in our family, and I was so crazy about him that my mother was fond of saying that my first words when I learned to talk were not either the traditional *ma ma* or *da da.* Instead they were *Dike Dike,* which was the best I could do with "Duke."

In any case, my mother was not my principal animal mentor. That honor belonged to an aunt of mine named Aunt Lu. The Boston in which I was born was very big on aunts. It was sort of the way elephant herds are—you know, the elephant aunts who help take care of the boy and girl elephants. Actually it seemed to me everybody in my Boston had aunts—indeed many more than seemed to have had uncles. They had especially more of what were called "maiden aunts." This was apparently because there were not enough uncles to go around—or, at least, not enough uncles who had gone around enough.

As for aunts, it seemed everybody had them. They either had regular aunts or great-aunts—aunts who were either the sister of their grandfather or grandmother—some of whom were maiden and some who were not. Why they were called great-aunts instead of grand-aunts I do not know, but they were. And some of them were either maiden great-aunts or great-maiden aunts, I do not remember which. I do remember, though, how many of them there were. I, for example, had at least half a dozen regular aunts and half a dozen maiden aunts, and, besides, at least three great regulars and two great maidens. And they were just the ones in my own family. Outside of my family I was familiar with a large number of other maiden aunts in many other families.

The most special of my aunts to me was a small, sharp, spry little lady who was close to my favorite person in the whole world. She had a face that seemed to wear a permanent quizzical look, although in those days I would not have known what that was. But even at a young age I did know that when she smiled, her smile was face-wide. Her name was Lucy but almost everybody it seemed, whether they were nieces or nephews or not, called her Aunt Lu.

I loved her for a lot of reasons, but the main reason was that at her house there were always dogs and cats, and I do not mean just a few dogs and cats—I mean lots of dogs and lots of cats. There were big dogs and little dogs and medium-sized dogs and odd-sized dogs. And, by the same token, there were big cats and little cats and medium-sized cats and odd-sized cats. And just as there were old dogs and puppy dogs, there were also old cats and kittens.

My aunt did not buy these animals—in fact, I cannot remember her ever buying any animal. She just found them or else, as I am sure happened many times, they found her. She was known all over her area—which was in Jamaica Plains, near Jamaica Pond—as the Stray Lady.

Nowadays there are, thank heaven, many thousands of Stray Ladies. But in those days, sadly, Stray Ladies were few and far between. A lot of people—most people, in fact—thought Aunt Lu was odd, and quite a number of them actually thought she was a little crazy. Some people even made fun of her. They told stories about how dirty her house was and how many messes the animals made, and they also talked about fights the animals had. From the beginning, even at a very young age, I didn't like those people. I loved her and her house and all the animals—I even loved the ones who occasionally bit me—and I hated people who made fun of them.

Aunt Lu was not young when I first knew her—she was in her sixties—and it was an awful lot of work for her to look after all those animals. She was not rich and, although she was not poor and could have afforded more help, she had a perennial shortage of such help because the servants she had did not enjoy having to look after all the animals and usually left after a short time. Indeed much of the time my Aunt Lu seemed to do all the cleaning alone but she never seemed to mind it at all, and as far as I could see, she was always cheerful.

One thing I could not help noticing about Aunt Lu and her animals was that she seemed to love them all equally—the pretty ones and the not-so-pretty ones, the nice, easy-to-love ones and the not-so-nice, not-easy-to-love ones. How was she able to do that? I wanted to know. She told me that she would not want any of the animals to think she did not like him or her as much as another. She also told me that being an animal lover was like being a mother—you had to like all your children, and the very worst thing you could do was to let any one of them, even the most difficult, think he or she was not loved. Finally, she told me that the longer she lived, the more she realized that the most difficult child or animal you had was very often, in the end, the one you loved best—if, and here she gave me that special quizzical look of hers, you allowed yourself to use the word *best,* which of course, she added, you shouldn't.

Another thing I noticed about Aunt Lu and her animals was that they almost never seemed to fight, despite what other people often said. Even the dogs and cats never seemed to fight. I noticed that she separated some from others in different parts of the house, but it was really remarkable how quickly she let them come back and join the others. I told her so. She thought I was exaggerating. "Oh," she told me, "they have their differences, and I'm sure they sometimes fight when I'm not around. But every single one of them knows the rules here, and the first rule is that I won't put up with fighting. I never punish them with anything except the way I speak to them, but if you speak to them a very special way when they are doing something wrong they learn to not do it."

I also remember well asking my Aunt Lu which were smarter—dogs or cats? "Well," she replied, "the great dog writer, Albert Payson Terhune, once set out to prove once and for all which was the smarter." She told me that what he did was to take a dog out for a long walk at the same time that a friend of his was also taking a cat out for a walk. Then when they both came back in and were tired and presumably thirsty, Mr. Terhune first put the dog in the bathroom in which a faucet was just trickling drop by drop, and then he took the cat in another bathroom where the faucet was similarly arranged. Both animals were given the same shot at the faucet. The dog was able to reach it on his own, but the cat was provided with steps.

In an extraordinarily short time the cat, by pushing and pulling at the handle of the faucet, was able to get a stream of water and a fine drink. But the dog never did master the problem. He lapped only a few drops and that was it. Of course, Aunt Lu pointed out that Mr. Terhune never meant by this that cats were smarter in all ways than dogs. In fact, he pointed out that the various areas in which dogs were smarter—Seeing Eye dogs, guard dogs, war dogs, tracking dogs, etc.—involved abilities cats apparently did not have. But he did point out that the main reason cats would not do those things was not necessarily that they could not do them—they just didn't want to do them. It was not part of their nature to be that dependent on what somebody else wanted them to do.

FRANKLY, AT THAT AGE IT DIDN'T MAKE ANY DIFFERence to me whether cats were smarter or not—I wanted a dog. Whether this was because of Duke or not, after Duke had passed on I wanted a dog and I wanted it to be my own dog. Though Aunt Lu took my side on the dog issue, even she could not prevail upon my mother. My mother felt that I had plenty of dogs over at Aunt Lu's and I did not need to have one at home and anyway, I wouldn't be around enough to look after it properly. My mother was a very tough customer when it came to arguing—she was the only girl in a family of seven brothers and apparently from her early life was used to getting her own way, or at least making it seem that nobody else got theirs over her disapproval. Actually the only reason I ever got a dog was pure luck and the fact that my grandmother, who used to come for Christmas every year, early in her visit overheard a conversation about the dog between my mother and me.

My grandmother always seemed to me half the size of my mother—she was really a very tiny woman. But if my mother had a mind of her own, where she got it of course was from my grandmother. In any case, in no time at all my grandmother was completely won over to my side, and from the moment she was, it was just a matter of time until I got my dog.

But it did take time. For one thing, my grandmother was not an easy person with whom to have a conversation, let alone have an argument. Not only was she formidable in the way she expressed her opinions, she was also formidable in her use of words. Talking with her was very much like doing a crossword puzzle—something at which, incidentally, she was extremely good. She would, for example, start a sentence and then, just when she was about to say the crucial word in that sentence, she would suddenly and completely stop. "You know," she would say, "your mother is, you know what I mean, you know the word I want," and then it would be up to you to supply the word. And, mind you, it did no good for you to come up with a word that was close to the meaning of the word she wanted—you had to come up with the *exact* word, or the whole conversation was still on hold.

And, make no mistake, if conversation was not easy with my grandmother, far harder was dissuading her from something she had her heart set on—like my getting a dog. My mother didn't have a chance, any more than I had a chance a few years later with an argument I had with my grandmother over Mahatma Gandhi.

When it came to people, my grandmother had many favorites and many unfavorites. And, first among all her favorites was Mahatma Gandhi. One morning when I was visiting her in New York, she saw something in the paper about Gandhi. She put down her paper and asked me if Milton Academy, the school I was attending, taught me about Gandhi. Immediately I knew I was in trouble. I knew well that Gandhi was a tremendous favorite of my grandmother's, and I also knew well that I would have difficulty satisfying her with an answer to her question. Lamely and keeping my voice down, all I could come up with was that he wore a towel and caused an awful lot of trouble to the British Empire. This time, when my grandmother spoke, I did not even have the hope of getting a crossword puzzle, which I would have preferred. Instead, extremely sternly, she said, "Is that all you know about Gandhi?" Miserably I nodded. With that my grandmother demanded to know when my Easter vacation—the time when I was visiting her—would be over. Knowing that I was making a terrible mistake, but also knowing I would have to bite the bullet, I named the day. My grandmother gave me the dreadful news—she would accompany me back to school.

As the day of our trip drew nearer I tried everything I could think of to make her change her mind, but nothing worked. Back to Boston she went with me in tow, and the next day, much to my chagrin, embarrassment, and all sorts of other rotten worries, off we went to Milton Academy. I do not remember everything that happened, but I remember enough to put the pieces together. My grandmother went first to the Headmaster, then to our History teacher, then to our Current Events teacher—here I lost track of her, but I was sure she went to a wide variety of other teachers as well. In any case, from that time on I am not exaggerating when I tell you that Milton Academy gave us something about Gandhi in every single subject—not excluding Latin and Math.

AS FOR THE DOG, AS I HAVE SAID, WITH MY GRANDmother on my side, and aided firmly by my Aunt Lu, my mother had no choice at all. The very next Christmas my grandmother brought to Boston a catalog of dog pictures and told me to pick out one I liked. I chose one of an Olde English Sheepdog and sure enough, on Valentine's Day, with love from my grandmother, my Olde English Sheepdog arrived.

I could hardly forget my first sight of him. For one thing, he arrived on the back of a truck in a crate with slats up the sides and, for what seemed to an eight-year-old an eternity, I couldn't really see him at all. All I could do was catch glimpses of some gray and white fur, moving restlessly around.

Finally, though, the man who had brought my dog climbed up on the back of the truck and, with a hammer and a screwdriver, began to tear the slats off one by one.

When he had pried off about three of these, I suddenly caught more than just glimpses—I saw my whole dog. But before I could register what I had seen, I was flat on the ground with a gray-and-white monster on top of my chest, pawing at me and slobbering on me. What had happened was that Brookie—i.e., the name I had decided on for my dog, also picked out from that catalog—had simply leaped over what remained of the crate, lit lightly on the back of the truck, and finally and heavily, not only into but onto me.

As I lay on my back, gasping both with excitement and for breath, I thought about something Aunt Lu had read me about Olde English Sheepdogs—that whatever any other puppy does, an Olde English Sheepdog also does, only more so. In any case, it was certainly true that at that moment, he was doing more so—and in spades. When at last I got a real look at him I could see that he was something halfway between a Teddy Bear and a Shaggy Dog story—although of course I didn't know then what a Shaggy Dog story was. But I would soon know that, whatever it was, Brookie would do more of that, too.

The only thing I didn't see at first about my dog was his eyes. He just didn't seem to have any. When I finally saw that he did have eyes, even if they were almost wholly hidden by his furry face, I remember that Aunt Lu had told me that although he didn't look as though he could see, he actually could. She also told me, however, that Olde English Sheepdogs made up for the fact they couldn't see too well by having incredible hearing—in fact, she told me that they had the best hearing of any animals except eels. That statement fascinated me so much that during the first days I had Brookie I searched high and low to find an eel. I wanted first to whisper something very low into Brookie's ear, until he couldn't hear me, and then I would whisper even lower into the eel's to see if he could. My whole plan never worked, however, because I never found an eel, even when I got so cross I tried shouting for one.

There was something else that I noticed that Brookie didn't have, and that was a tail. Aunt Lu told me that the reason Olde English Sheepdogs didn't have tails was because in the old days in England they had a tax on hunting dogs, and although Olde English Sheepdogs were herding dogs, they were not hunting dogs, and so their tails were cut off so they wouldn't be mistaken as hunting dogs and be subject to the tax.

I told her I thought it was cruel—that all dogs should have tails. Aunt Lu not only agreed with me, she also told me that when people cut off the tails of newborn puppies it hurts so much, and the puppies scream so much, that when it was being done their mothers had to be taken away so they couldn't come to stop their puppies from being hurt.

The next thing I wanted to know was why the tail cropping was still done when the tax on hunting dogs didn't exist anymore—although frankly I would have liked it if it still did. The reason, Aunt Lu told me, was that it was tradition. If there was anything that kennel clubs really liked, it was keeping up tradition.

As time went by Aunt Lu told me many other things about Olde English Sheepdogs—not the least of which was how they had become so popular. In fact, they became the most popular dog in the shortest time of any breed in history. One way they did this was by having the terrific pioneer breeder on their side—none other than Andrew Carnegie himself. Apparently Mr. Carnegie not only had many Sheepdogs himself but also got many of his friends to acquire them. Indeed, he had one friend who was so enamored of Olde English Sheepdogs that whenever he had a mother Sheepdog who had puppies, he would take them over, one at a time, to his friends' houses and leave them. He liked to say he never had a friend who turned him down.

Even at the age of eight I thought this was a pretty terrible way to give away puppies, but I had to admit I wasn't surprised. I loved everything about Brookie—even brushing and combing him which was, frankly, an endless job. More than one showperson has told me that to put an Olde English Sheepdog in shape—bathing, combing, and brushing—for a show could take a whole day. Needless to say, I never got Brookie into that kind of shape, but I was so proud of the way I did get him looking that the minute the spring vacation was over I decided to take him to school. But I really didn't do it just to show him off—I did it because, although I had only been with him a couple of weeks, I couldn't bear to be away from him.

The school was a small, private school not far from my home—just down a long path. The moment I got there everybody was so excited that they made Brookie excited, too, and I couldn't control him very well, even though I had a leash. Finally, though, the bell rang and I took him as quickly and quietly as I could over to my desk and shoved him as far underneath as I could get him. I was not quite either quick or quiet enough, however, to foil the teacher. Fixing a sharp eye first on Brookie and then on me, she said, "What is that bear doing under your desk?"

I protested it was not a bear; it was an Olde English Sheepdog. I told her it was pronounced "Old," even though it was spelled "Olde." I thought that would please her, because she was a stickler for spelling, but unfortunately, it did not. "I do not care how that dog is spelled," she said firmly. "He does not belong in a schoolroom." Once more I tried to protest. I told her he doesn't mind it at all. I think he likes it. Once more it was her turn. "I do not care whether he likes it or not," she said. "I want you to take that dog home immediately." I pulled Brookie out from under the desk and stood up, but I had one more hope. If I take him home now, I asked her, may I bring him back at recess? I told her I thought he would be terrific at recess because we played Prisoner's Base, and since Brookie loved to herd people, I told her I thought he would love the game.

That teacher, I am sorry to say, was very closed-minded about Brookie. I had to take him home, and she wouldn't even let me bring him back for recess because she obviously didn't care whether he liked Prisoner's Base or not. In the end, however, Brookie and I won, because Brookie would hear the noise of us all going out at recess and he would come down on his own. And, just as I said, he was terrific at Prisoner's Base.

IF, AS I SAY, THE ARRIVAL OF MY DOG WAS THE MOST memorable moment of my childhood, what came in a close second was another event that was almost equally memorable, and that would last for the rest of my life—the arrival of, of all things, a book. It was a gift, and it was from Aunt Lu. And, since it was from her, it was, as you might guess, an animal book. But make no mistake: it was not just *an* animal book; it was *the* animal book—a book that is still to this day arguably the greatest animal book ever written.

The book was *Black Beauty*, and it became in short order my favorite book. Curiously neither Aunt Lu nor I was associated particularly with horses. I knew she probably liked them, the way she liked all animals, but it would never have occurred to me that she liked them the way, for example, she liked the dogs and cats she rescued. I was soon to learn, however, that she did. As for myself, up to that time I knew very little about horses. My father and mother went riding on Saturdays and Sundays at a nearby stable, and my brother and I and my sister, when we were old enough, went with them. Nonetheless, in the vernacular of the day, we were not a "horsey" family.

I did not have to wait long to realize how wrong I had been about Aunt Lu and horses in general and *Black Beauty* in particular. She loved the book so much it seemed that, whenever I told her where I was in it, she not only knew exactly where that was but could even tell me, word for word, what went on there. And I could hardly help realizing that, more important than how she felt about it herself, she obviously wanted me to know the book, if not by heart as she did, at least in my heart.

Under such tutelage it was not surprising that I soon did. Aunt Lu made clear to me that the incredible achievement of *Black Beauty* was that the story was so engrossing that on one level it can be read simply for enjoyment and, on another, for excitement—in such scenes as the fire and the rescue. But, on another level entirely, it can be read as a complete manual of man's inhumanity toward horse. "Remember," I can hear my Aunt Lu saying, "you are not reading a children's book. It is not a children's book because, although it is a book for children, it is a book for adults as well—indeed, one for both children and adults and for that matter for anybody who does anything mean to horses or to other animals."

I suppose I loved best the very first part where Black Beauty is growing up with his mother and the other colts under a master who, as Black Beauty says, "spoke as kindly to us as he did to his little children." But even as a child I was soon to learn that all was not going to be sweetness and light in my new book. The very second chapter is called, for example, "The Hunt," and although Black Beauty is too young to have a part in it, Squire's son is killed in the hunt, and Black Beauty's brother breaks his leg and has to be shot. " 'Twas all," the author tells us, "over a little hare." And how does that little hare do? She tries to get through a fence, but it is too thick. The dogs were upon her, with their wild cries: "We heard one shriek, and that was the end of her."

ONE DAY WHEN I WAS READING MY NEW BOOK ABsorbedly, my Aunt Lu sat down and talked with me very earnestly. She told me she wanted me to know several things about the book *Black Beauty*. The first one, she said, was that it was not just a book *about* a horse; it was a book about a horse *by* a horse, an autobiography of a horse. The second was the author had written it not just with sympathy for the horse but something much more than sympathy—something even beyond empathy. She had put herself literally into the mind of the horse.

The third and most important thing Aunt Lu told me that she wanted me to know about the book was that the man who was responsible for bringing it to America—it had, of course, first been published in England—was not only an uncle of hers; he was also a great-uncle of mine. His name, she said, was George Thorndike Angell.

It had all come about, Aunt Lu explained, because of something that had happened back in 1868, ten years before the publication of *Black Beauty*. It was something, she said, that had made my great-uncle terribly upset. What it was was a long-distance horse race. Aunt Lu explained to me that in those days long-distance horse races were very common and they were almost always cruel but this one, she said, was particularly cruel. It was a race between two famous trotting horses named Empire State and Ivanhoe for a purse of $1,000. It was a forty-mile race, and both horses averaged a speed of more than fifteen miles an hour and both, after the race, died. They had literally been driven to death.

The very next day my great-uncle wrote a letter to a Boston newspaper saying that he was willing, as he put it, to contribute both time and money to stop such cruelty to animals and, appealing to others, as he put it in the fashion of the day, "with whom I can unite and who will unite with me in this matter."

Stiff as that letter may appear to us today, an extraordinary number of people did write to "unite" with him. Boston had long had a large number of prominent people who were willing to speak out in those days against cruelty to animals. They included, among others, Ralph Waldo Emerson, Henry David Thoreau, Dr. Oliver Wendell Holmes, and John Greenleaf Whittier. My great-uncle was not as well known as these, but he was a formidable man nonetheless. Tall and handsome, with a Lincolnesque look, he was the son of a minister who had died when my great-uncle was a child, leaving him and his mother in severe financial straits. He became by his own abilities one of Boston's outstanding lawyers and philanthropists. He stayed a philanthropist, but he did not stay a lawyer. The trotting race had so infuriated him that he not only virtually abandoned his law practice but also was soon devoting almost all his time to stopping cruelty to animals. Soon, with hundreds of others, he formed the American Humane Education Society, the first of its kind in America.

Aunt Lu told me that even before the horse race my great-uncle had wanted to have a book written that in his opinion would "do for horses what *Uncle Tom's Cabin* had done for slaves." He enlisted all kinds of American writers to do the job, but either none would do it or those who were willing to do it did not satisfy him. Finally, out of a clear sky one day came a book mailed to him by a Miss Georgiana Kendall from New York City. My great-uncle did not know Miss Kendall, but she clearly knew of him.

The book was *Black Beauty*. My great-uncle apparently read it at one sitting, and from the moment he finished it he went to work. In his own publication of the American

Humane Education Society, *Our Dumb Animals,* he wrote as follows:

> *I want to print immediately* a hundred thousand *copies here.*
>
> *I want the power* to give away thousands of these to drivers of horses—*and in public schools—and elsewhere.*
>
> *I want to send a copy, postpaid, to the editors of each of about* thirteen thousand *American newspapers and magazines.*

In just two months the first American edition of *Black Beauty* was published, the actual date being April 1, 1890—a date that we may be sure was made fun of by some of my great-uncle's antianimal friends. In any case, by the end of the year the edition had sold 216,000 copies and my great-uncle had given away another 216,000. The title cover read *Black Beauty; His Grooms and Companions, by A. Sewell—The Uncle Tom's Cabin of the Horse.* The author's name being just "A. Sewell" instead of her whole name, Anna Sewell, was something that had obviously been done on purpose because the publishers of the book thought that a book on animal cruelty would attract more attention if the author was perceived to be a man rather than a woman. One thing was certain—attention it did attract. Angell's giveaways, particularly to drivers of horses, were an important factor in drawing this attention and led to widening sales not only in America but in England as well. One woman philanthropist in England, for example, did her best to give a copy to every single driver of a horse-drawn railway van.

Nor did my great-uncle count on English people alone to do the job in that country. A wide traveler, he made many trips to England and, among other things, went to a meeting at which he asked the English to start a sister publication of *Our Dumb Animals* entitled *Animal World*. Following the three-hour meeting, my great-uncle noted in his autobiography, something happened that he never forgot:

> *One thing at the close struck me as very strange, and that was the question,* Who should move the vote of thanks? *It seemed to be regarded as a much more important matter than we consider it in America, but presently Field-Marshal Sir John Burgoyne, an aged gentleman, very near the head of the British army, slowly rose, and with the utmost dignity moved the vote of thanks.*

The fact that my great-uncle was thanked by the son of the famous English general who surrendered to George Washington was fascinating to me. But my Aunt Lu would not let me go overboard on him. "Your great-uncle was no saint," she told me. "Like all Thorndikes and Amorys," she continued, "he had his quirks." And here she fixed me with her quizzical look.

I didn't know what quirks were, but I wanted to and so I asked her. She told me they were idiosyncrasies, which certainly didn't help me much so I gave up. She went on to say my great-uncle had three other Causes besides animals. One was against adulteration of foodstuffs; another was against poisoning in wallpaper manufacturing; and still a third was his concern about people getting buried alive. From earliest childhood, Aunt Lu told me, he himself feared that he might be buried alive, and he particularly favored the habit, not only of Bavarians but also of some Victorians, of burying a person with a bell attached to one of his or her big toes.

That quirk, or idiosyncrasy, whatever it was, really did intrigue me. I'd never heard of such a thing, and it struck me as so funny that I could not wait to get my brother and go visit a cemetery to see if we could hear any bells ringing. We decided to take Brookie with us. This was a mistake because at the cemetery we visited there was a burial service going on and Brookie, who had never seen a burial before, got so interested in it that he started barking. At this people came along and pushed us away and even tried to shoo Brookie off, which we thought was very mean of them because it wasn't his fault—he wasn't going to do anything about a bell, whether they put one on somebody or not.

We never did get to find out what we wanted to in the cemetery but, quirks or idiosyncrasies or not, my great-uncle's work with *Black Beauty* alone would have established him as by all odds one of the greatest humanitarians of the nineteenth century. By the end of his life he had given away more than a million copies of the book to libraries, schools, stables, and the public. Nor did his work stop there. He would also achieve another feat—the publication of *Beautiful Joe,* a book about cruelty to dogs which was written by a Canadian author and which he regarded as close to being on a par with *Black Beauty.*

BESIDES TELLING ME ABOUT MY GREAT-UNCLE AND THE crucial part he played in the enormous success of *Black Beauty,* Aunt Lu spent much time talking about the book's author. It was a sad story. Anna Sewell came from Quaker stock. Her family was a close-knit one, consisting of a not-too-successful father who had many different jobs and a mother who was an extremely dominating figure, as well as a brother who became an engineer and lived a large part of his life abroad. The only known portrait of Anna, painted when she was sixteen, shows not a beautiful girl but an interesting-looking one with a round, loving face and large penetrating brown eyes. A bright but extraordinarily gentle person, she was called in the fashion of the time "sweet tempered."

At the age of eleven she suffered a curious accident. Returning from school, she was caught in a heavy rainstorm and, having no umbrella, ran home. The road sloped steeply in front of her gate and, just as she reached the gate, she slipped and fell. It was a heavy fall, and she badly injured her ankle. Although at first the accident was regarded as just a bad sprain, it was not well treated. It made her lame, not just in one leg but also in the other, and from that time on she not only never walked without difficulty and pain, she could not even stand without leaning on something for support, let alone walk any distance.

As a result of her disability, since she was so limited in walking, her early life revolved around either riding a horse, when her legs permitted it, or driving one. As a child driver she was known as "a very good whip"—an expression that in her case was an irony in the extreme, because not only did she never use a whip in her life, most of the time she did not even appear to use the reins. She relied instead on her voice. Her favorite task was driving her father to the station in her pony cart—down to the station in the morning and back in the evening to meet him and drive him home. Her mother's biographer gives a description:

> *She seemed simply to hold the reins in her hands, trusting to her voice to give all needed directions to her horse. She evidently believed in a horse having a moral nature, if we may judge by her mode of remonstrance . . . "Now thee must go a little faster—thee would be sorry for us to be late at the station."*

Besides the attention to her animal in riding or driving, Anna spent endless time in the stable. Here she would not only talk to the horse or pony, she would also see for herself that the water bucket was full, the straw fresh and clean, the animal's coat thoroughly brushed, and his hooves picked out. "On warm Sundays," *Black Beauty* historian Susan Chitty has told us, "she would have been sure to give orders that he be turned into the paddock for the day."

In addition to loving horses, Anna loved language. Her mother was a poet, and one day they received extraordinary news. A publisher had taken a fancy to Anna's mother's verses and was going to publish them. Today these verses seem far from deathless. One, for example, was entitled "Mother's Last Words." It was about two orphan boys who, after their mother's death, apparently never forgot her dying words instructing them always to be good and never to steal. The older boy nonetheless does steal—a pair of boots for his younger brother—but he is so overcome with remorse that he returns them. Whereupon, the ballad tells us, the younger brother promptly dies of cold. But everything turns out all right in the end, at least by Anna's mother's standards, because the older brother evidently never again sins and, as for the younger one, he may be dead but he has, as the ballad tells us, "found, once more, his mother dear."

The "Mother's Last Words" volume sold, my Aunt Lu told me, a million copies—something I had found, even at that age, as difficult to believe as anything she had ever told me. But what my Aunt Lu was most interested in was that, throughout Anna's mother's successful writing career, her sternest critic had been Anna herself. Frankly, I would have thought she had no critic but "My Nanny," as she called her daughter, "has always been my critic and counselor," she said. "I have never made a plan for anything without submitting it to her judgment. Every line I have written has been at her feet, before it has gone forth to the world. If I can only pass my Nanny, I don't fear the world after that."

At the same time the mother-daughter relationship, Victorian as it was, could have its moments of fun. Often, Mrs. Sewell's biographer notes, the two played word games that involved putting words, within a time frame, into a rhymed ballad. In one of these Anna was given six words—*prawn, yawn, tall, wall, missed,* and *kissed.* And, within the time frame allowed, she came up with her ballad:

O Henry dear, don't yawn so loud,
The tea will soon be here;
But Jane has had an accident
Which might have cost her dear.

In coming in with that tall urn,
She caught a sight of Jem,
She missed the step, and kissed the wall,
Instead of kissing him.

I do not think the girl is hurt.
But still she's vexed and fluttered;
So cook will bring the prawns and toast,
And tea-cake, when 'tis buttered.

Finally, in the early 1870s, at long last, and just seven years before her death, Anna herself began to write. There are many different opinions on how sick she was by the time she began *Black Beauty*. Two things are certain—one is, at the same time she began the book, she also contracted some kind of a mortal disease—one that in those days very little was known about. The other is that most of the time she was writing the book she was too ill to hold a pen. Indeed, the greater part of *Black*

Beauty seems to have been first memorized by Anna and then passed on to her mother on little penciled slips of paper which her mother would then copy. Susan Chitty also maintained that Anna had a way of being able to think out a whole chapter, go over it and over it in her mind, and then literally recite it word for word to her mother. Again, as my Aunt Lu liked to point out, Anna's own aunt played no small part in coaxing Anna to commit *Black Beauty* to paper. By 1876, in Anna's own brief diary, there is concrete proof in the entry of December 6: "I am getting on," she wrote, "with my little book *Black Beauty*."

It was her mother who finally took the manuscript to the London manager of the publisher who had for so long published her works. As Mrs. Chitty recalls, Mrs. Sewell asked the manager to give her an opinion on "this little thing of my daughter's," and left the manuscript with him. A few days later a letter arrived from the manager offering an outright payment of twenty pounds for the copyright of *Black Beauty*. Her mother advised her daughter to accept, pointing out that an unknown author of a first book could not expect more.

On August 21, 1877, Anna made her third and final

entry in her diary about *Black Beauty*. "My proofs of *Black Beauty* are come," she wrote. "Very nice type." The book was published on November 24, 1877, and on April 23, 1878, Anna Sewell died.

The funeral was a simple one. As the small funeral procession passed the Sewell house, Mrs. Sewell, before joining it, happened to look out her window and saw that the team of horses that would carry Anna's coffin to the burial ground had, on their heads, check reins—something that, of all pieces of harness, Anna detested the most. "Oh no," Mrs. Sewell exclaimed, "this will never do," as she hurried down to the cortege.

Shortly afterwards, all the guests observed a moving scene. The driver of the cortege, a top-hatted gentleman, went slowly down the line of horses, removing the check reins of each horse in turn.

IT WAS NOT LONG AFTER READING *BLACK BEAUTY* FOR the first of many times that I had a dream that one day I would have a place which would embody everything Black Beauty loved about his final home. I dreamed that I would go even a step further—at my place none of the horses would ever wear a bit or blinkers or check reins, or in fact have any reins at all, because they would never pull a cart, a carriage, a cab, or anything else. Indeed, they would never even be ridden—they would just run free.

I even dreamed about the sign that would be on the gate at the ranch. It was certainly not hard to imagine this, because it would have on it, I decided, the very same words as the last lines of *Black Beauty*.

I have nothing to fear,
and here my story ends.
My troubles are all over,
and I am at home.

Today, I have often wished my Aunt Lu had lived to see my childhood dream come true. I think she would especially have liked to see that the sign, with *Black Beauty*'s last lines on it, not only read "Last Lines of *Black Beauty*" but also did not just have "A. Sewell" on it but "Anna Sewell."

My Aunt Lu was indeed right. Although Miss Sewell received only twenty pounds for a book that became the sixth-largest-selling book in the English language, and although she died just a few months after its publication and never lived to see its success, she even became relatively anonymous in death, since her little burial ground was soon destroyed. But none of these things would have mattered to her, as anyone who knew her would say. What would have mattered, and would have given her all the satisfaction her modest self would have wished, was that her book would do more than any other book ever written—not just for horses but, arguably, for all animals.

Chapter Two

OF GOOD OLD BOYS, A THREE-LEGGED CAT, AND A BURRO NAMED FRIENDLY

Even in my dreams, I apparently thought more about what the Black Beauty Ranch would not be rather than about what it would be. For one thing, I thought firmly, it would not be a zoo. It would have wild animals, but it would not be a place where the wild animals were there primarily to be looked at; rather it would be a place where they would primarily be looked after. It would also not be a place where people ate popcorn and perhaps even gave it to the animals, and it would certainly not be a place where people thought about how funny the animals looked. For still another thing, it would not be a farming ranch where animals were there primarily to be eventually eaten. And, for the next-to-last thing, it would not be a racing ranch or a rodeo ranch or any other kind of a ranch where animals had to compete against one another. And for the final thing, my ranch would most definitely not be a place for circus acts. No animal would stand on two legs or sit on a stool or jump through hoops or do tricks or acts or any other kind of stunt.

What my ranch would be, on the other hand, I saw very clearly in my dreams. It would be a place where animals would do whatever they wanted to do, not what people wanted them to do, and particularly not what people wanted them to do when they were watching them. It would also be a place that the animals felt, from the day they arrived, belonged to them, and would always belong to them as long as they lived.

The only thing I did not have any idea of in my dreams was where all this was going to be. In the end, I chose a place that was just about as curious a place for a Bostonian to choose as could be imagined. It was, in fact, none other than Texas, and I cannot excuse this choice by saying I did not know anything about Texas when I picked it. I had been to Texas many times as a visitor and had written several magazine articles about it. I had even been told that Texas would be the very best place for what I seemed to be dreaming about because it was one of the best places in the whole country for good grass-lands for animals. At the time I heard this I did not know anything about grasslands, and I am even pretty poor when it comes to that knowledge still. I have, however, learned that Bermuda grass, or as Texans pronounce it, "Bermooda grass," is indeed an almost ideal grazing grass for most hooved animals.

I also did not then know just how many drawbacks there would be to Texas. I was, however, soon to find out. In the first place, consider the climate. J. Frank Dobie, the great Texas writer, once said that Texas did not have a climate; it just had weather. And, if any other state ever had such weather it is one in which few men alive, and presumably an ever larger number dead, were ever privi-leged to live. Not to beat about the bush—which in Texas is all too easy to do for almost any reason but par-ticularly out of frustration—Texas is as cold as an ice-berg in the winter and as hot as a furnace in the summer. In between, in the fall, you have freezing winds which

Texans, who are not fond of anything from the North, call "Blue Northers." Then in the spring you have on the one hand literally dozens of tornadoes and, on the other, a quaint combination of two of the most dreadful plagues anyone can imagine—fire ants of indeterminate size and mosquitoes so large it has been said of them that they require runways on which to land. On top of having a wide variety of other plagues—including, but not limited to, locusts—Texas also has a long history of almost incredibly long droughts—so many indeed and so long that it is a much labored Texas joke that when, in the time of Noah, it rained for forty days and forty nights Texas got exactly half an inch.

All in all the question of climate in Texas was perhaps best summed up by General Phil Sheridan, who spent some time there after the late Unpleasantness Between the States. "If I owned Texas and Hell," he said, "I would rent out Texas and live in Hell."

In the second place, consider the scenery. Texas is not the Sahara Desert, but it is in large part close enough to it so that it can hardly be denied that it is a place where you can see farther and see less than any place on Earth save possibly for a place in Texas you haven't seen yet.

The vast majority of the land is flat as a pancake, and when Texans talk about plains do not for a moment deceive yourself that they are talking about any great plains—they are, rather, talking about extremely plain plains. It is true that there is an abundance of cactus around and it is also true that there are many people who are apparently blessed with being able to see great beauty in cactus, but to do so you apparently not only have to be blessed but you also have to have lived in Texas for no less than a generation and a half—something that in Texas qualifies you, if not for eternity, at least for being an awfully good old boy.

All in all if Texas was, generally speaking, totally deficient in terms of a reasonably decent climate or decent scenery, it also seemed—even worse from the point of view of my childhood dream—deficient in terms of having a reasonably decent attitude toward animals. Indeed, before I knew much about Texas, I firmly believed that all Texans thought that animals were good for just three things—to make money off of, to eat, and to shoot. Even pets in Texas, it then seemed to me, appeared to be pets only to a limited extent—one that did not include any relief from being either chained up or closely confined if they were indoors, or, if they were outdoors, riding in the back of a pickup truck with very little hope of doing anything but, sooner or later, falling off.

There was, however, something that made the location of the ranch far from undesirable. Contrary to so much of Texas, which, as we have pointed out, had so little, East Texas—the area in which we bought the land—was something very different. Almost from the moment when you leave the Dallas airport and journey southeast toward Shreveport in Louisiana, the grass, mercifully enough, grows steadily greener, the land equally mercifully grows steadily less flat, and there is, most mercifully of all, more and more water around. Indeed there are so many lakes and ponds and rivers that the first time I saw Murchison, Texas, the post office town between Athens and Tyler, I had to rub my eyes to believe I was not in New England.

I know this will sound difficult for people to believe, but it is true—the first acres we bought had two lakes on them, and since that time, when the Ranch has grown to a thousand acres, it has no fewer than four lakes, let alone a dozen ponds and brooks, and makes the entire location, compared with the ideas I'd had about this part of

Texas, if not Paradise, at least not so far from this side of it.

At this point you may well ask if we found the people in East Texas, as we grew to know them, particularly different from Texans elsewhere in the state. It would be nice to say yes if for no other reason than that we do, after all, have to live there. But that answer would be too easy, because the fact is as we grew to know them we found that not in East Texas or anywhere else in the state do Texans consider cruelty to animals as a way of life. Indeed we soon found that there were just as many people who cared about animals and even had empathy toward them as there are in other states. And I do not mean by this just people with whom we worked or who helped us at the Ranch—they would of course feel that way or they wouldn't be there. I mean Texans in general.

And this brings up another point. When it comes to the character of Texans in general, to give the devils their due as it is sometimes necessary to do, it is only fair to remember where Texans came from. They were, to begin with, not called Texans; they were called "Texians," and they were also, as they never let an outsider forget, once a whole nation. It was a messy nation, to be sure—one originally settled by drifters and gamblers and bandits—but among these early settlers, it should also not be forgotten, were some 185 defenders who for thirteen days fought to the last man against some 4,000 Mexicans at the Alamo and not too long thereafter, under the able Sam Houston, were able to avenge the Alamo and completely defeat the Mexicans at the battle of San Jacinto. Even after that, however, they were still not out of the woods. This was because, when their hero Houston became their governor and wanted them to be annexed by the United States, they not only refused to do so but also for good measure threw him out of office. When eventually they finally did agree to be annexed and become a state they did so only on the condition that the United States would not get any of Texans' public lands. To this day, Texas is the only state in the union that, with the exception of one national park, owns all its public lands.

As if this were not enough, by the time of its annexation Texas had the right, and apparently still continues to have such a right, to divide itself, if it so chooses, into five separate states. From the beginning, however, it clearly showed no disposition to do any such thing. Indeed when the governor who succeeded Houston, a man named Mirabeau B. Lamar, took office he immediately showed he was a Texan through and through. For one thing, he was so nervous during his inauguration that he was unable to deliver his inaugural address but instead shuffled about while it was delivered by his secretary. For another, he soon showed that not only did he not have the slightest intention of dividing Texas, he instead did the opposite and embarked upon a military campaign in which he intended to enlarge Texas all the way to the Pacific Ocean. Happily for all of us, he failed.

WE DID NOT FAIL, HOWEVER, IN OUR PERSISTENCE TO establish Black Beauty Ranch in Texas. The first customer was a young kitten—a kitten who is today the Ranch's senior cat citizen. No one who saw her the day she arrived will ever forget the sight. She was crawling toward the main house, step by painful step, all the time caught in and dragging a leghold trap. Where she had been caught, and how far she had come, we would never know. We did know, however, that she had come to where she hoped to get help from the jaws that held her in their viselike, grisly grip.

She had hoped right. We immediately took her to the nearest veterinarian. His diagnosis was sad. She would have to have removed not only her front paw but also her whole leg, all the way up to her shoulder. She also had to stay at the vet's for some time. When we were finally able to get her back, we could hardly bear watching her as she began what would be her lifelong necessity of trying to get along on a three-pawed hop.

We named her Peg—never to be used with even the hint of a peg as in "peg legged." Instead, it was always Peg as in "Peg o' My Heart" because from the very beginning she not only became the first customer of the Ranch in point of time, she also became, in all our hearts, first of all the animals.

The main reason was, of course, her cruel disability. Front legs are such a vital part of almost everything a cat does, except possibly jumping—and even partly in that maneuver—so that every time she moved it seemed to hurt us watching her far more, thank heaven, than it seemed to bother her. One thing, however, from the beginning that we loved to watch her do was wash her face. We loved this because, as time went on, she also learned a new maneuver for this important activity—twisting the one front paw she did have around and very cleverly backhanding it to the business at hand. This was by no means the only clever thing she had to relearn; it was just one of many.

Today as I write these words Peg has been with us for eighteen full years, and neither we nor any one of the vets we have taken her to has ever been able to be exactly sure how old she is. That is because, of course, we don't know how old she was when she came to us. Cats are not like horses—you cannot tell by looking at their teeth, which in any case is something that should be done, for more than one reason, sparingly—and cats' ages are difficult, even for experts, to be sure of. But one thing was certain—if Peg at first was the only animal at the Ranch, in a very short time she reigned over a very large number, not the least of whom were other three-leggeders.

Make no mistake: there have been plenty of these over the years. There is, for example, a three-legged raccoon, still another trap victim, whom we named ThreePete—following his friend RePete, a two-time rehab customer—who, on his third attempt, and on his three strong legs, was finally able to make it to be released in the wildlife area of the ranch. And besides there were three three-legged foxes, two of whom eventually made it to the wildlife area but one, Cassidy, named of course for Hopalong, had been so badly abused at a roadside zoo, where he actually lost his front leg, that we kept him near the Ranch house. Because not only was he not even a little afraid of people—an unfortunate prerequisite for releasing wild animals—but he thrived in our company and in the company of the other nearby animals. In other words, he would have made a terrible release candidate for the wildlife area, and so we kept him as a pet. As for the several three-legged coyotes, one lost a leg to a farm vehicle so badly she could not make it to the wildlife area but two others, including a coyote named Cool Leg Luke, did make it.

Finally, we had for many years a deer named Dixie. When she was a few days old she was run over by a hay sickle when she had, fawn-fashion, simply lain down as low as she could while the hay sickle went over her. This instinctive defense strategy, often risky, saved her life but cost her a leg. The farmer who was driving, however, got Dixie first to a vet and then to the Black Beauty Ranch where, after months of practice, she learned to run just like the other deer, and from a distance you could not even tell them apart.

Besides her job with all these three-leggeders, Peg is also the ruler of, and keeping a weather eye on, two resident dogs—Jake, who is also three-legged, and Shorty, who is so lame she might as well be. Peg too has charge of a number of barn cats, all of whom are, of course, neutered and spayed and well fed, and serve as the Ranch's feral colony. Then, too, there is the matter of another Ranch house cat named Polaris. This cat came to us in a curious way. We had all been to Dallas to promote the opening of a spay/neuter program at the local shelter, and during the opening the press was asked to view all the animals who later that day would have to be put down because they had found no adopters. During this sad viewing I spotted one white cat who reminded me of my cat Polar Bear, the stray who had adopted me so many years before and had inspired me to write *The Cat Who Came for Christmas.* In fact, I could not resist going over to his cage and playing with him.

I thought of him, of course, that night. And then, to my amazement, the next morning, all the way from Dallas to the Ranch, which is something of a distance, two of the organizers of the shelter visit came to the door. With them they had a cat. As I went out to greet them I saw immediately that they had that very white cat I had played with before what I thought would be his end. He still reminded me so much of Polar Bear that I wanted a "Polar" name for him, but I was fresh out of "Polar" names. We already had a Polar Star at the New York office. But in no time at all—resourceful in emergency as I always am—I christened the new cat Polaris. He has turned out to be second in command only to Peg as boss of an extraordinary number of animals.

Polar Bear, who is buried at the Ranch, did have one visit to the Ranch during his lifetime. And from the moment he and Peg met, there was an understanding between them I had never seen with Polar Bear and any other animal. Perhaps it was the fact that after a day at the Ranch spent with so many other animals he was at last face-to-face with one of his own kind. Or it could have been that he seemed immediately to recognize Peg's infirmity and not only respected it, but also felt he would not be offered the kind of challenge he might have expected from barging into another cat's domain.

In any case what happened was extraordinary. Almost immediately upon seeing Peg, Polar Bear leaped up onto the sofa and went into his Buddha meditation pose—

gazing directly at Peg, seated across the way. Then, slowly, Peg arose, looked around for a moment, and then hopped over toward Polar Bear. Just before she hopped toward him, however, she turned a little sideways and so landed not on him but beside him. As Polar Bear turned to look at her she in turn turned to look at him, but neither look from either one toward the other betrayed even the slightest hint of the typical hostile cat stare.

And, not immediately but in good cat time, Polar Bear slowly turned over on his side and reached out with all four paws toward Peg. So too, again in cat time, did Peg reach out toward Polar Bear—albeit with just three paws. In just a moment all seven paws were actually touching each other's stomachs and in that position first Polar Bear, who had after all that first day at the Ranch had a very long day, closed his eyes. At that Peg too closed hers, and both were soon asleep.

For some time I just sat and looked at my two favorite cats, together for the first time. I looked of course at Polar Bear for a long time, but I also looked at Peg. To us at the Ranch Peg has long been our living symbol of the cruelty of the fur business. No animal, with the exceptions of the monkey, the raccoon, and perhaps the otter, uses his or her frontal extremities more deftly than the cat. To go through the rest of her life without the use of a front paw was tragic enough, but in Peg's case, remember, it was not just a front paw, it was a whole front leg.

WE HAD NOT, OF COURSE, ACQUIRED THE RANCH AS A home for cats and dogs. We had indeed acquired it primarily to be, first and foremost, a home for—of all animals—burros. The reason for this was something that happened some distance from Texas, in the State of Arizona, in the Grand Canyon. What had happened was that the Grand Canyon National Park Service had, in its finite wisdom, declared war on the burros in the Canyon. In fact they had decided, on information conveyed to us by Richard Negus, the Fund for Animals' Arizona correspondent, to shoot the burros.

At that time we had no definite plans about what to do about their plan, except one. And that was that if the Grand Canyon National Park was going to declare war on their burros, then we were going to declare war on the Grand Canyon National Park. As a first step, we decided to sue the Park Service. We knew we had very little chance of success in this, but our plan had at least the virtue of being able to give us time, in the event of failure, to plan what our next step would be. This, we shortly but firmly decided, would be to try to rescue the burros. And, before we had any idea of how to do this, we did one very practical thing—we put down our first payment on the Black Beauty Ranch. After that we knew that if we ever did get to rescue the burros, we would have a place to put them.

When I said that the Park people were going to shoot the burros, I am not talking about the mules that people ride when they go down into the Grand Canyon. I am talking about burros or, as they are also called, donkeys—which, incidentally, is a word they dislike almost as much as they dislike the other name that they are also called, which is better known for being the backside of people than it is for anything having to do with an animal and which we will not dignify by even mentioning. In any case, the Grand Canyon people called their burros "wild burros," which I suppose was their privilege since the burros lived in their Canyon. Over the years, however, I have become very suspicious of a lot of animals that are called "wild" as if they were wild in the sense of being fierce, when in reality the only reason they are wild is that they have no home with people. They are not wild; they are feral, and they have returned to be wild only because they had to. Anyway, then and there we decided, before we began our war with the Park people, that if we did not win our suit, we would go for the possibility of rescuing the burros and afterwards adopting them out as, if not pets, at least companion animals.

As the Park people pointed out to us, no previous group who had ever tried to get burros out of the Park—and there were, incidentally, very few of these—had ever had much success. A few animals here and there had been taken out by literally being pulled and pushed up the long trail, but only one group had ever come close to succeeding with many burros. This was a group that used not only people on horseback but also dogs to pursue the burros. And, actually, they did get within sight of their objective—the top of the Canyon. But when the burros too saw the objective, they simply turned around and bolted past both the people on horseback and the dogs and went all the way down to the bottom—from whence they had come.

That bottom, incidentally, was seven thousand feet

down, and considering not only this fact, but also all the other failed rescues, we soon decided that we would not try either to pull the burros up the trail or to push them up by pursuit and pressure. Instead, we resolved to lift them up by helicopter, burro by burro and helicopter by helicopter, with a sling underneath the helicopter to carry each burro. We knew perfectly well that the whole thing would be a very expensive operation. Helicopter time does not come cheap, nor do helicopter pilots, particularly when you are, as we were, determined to get the best. Besides all this, we had to figure out a way of getting the helicopters to the burros or, conversely, the burros to the helicopters.

In the end we decided on roping them, and to do this I set out to find not just good ropers but, again, like the helicopter pilots, the best ones. It was an interesting search, particularly as it soon involved close attention to a sport I had long detested for its cruelty—rodeo. However, I soon realized that if we were not going to practice cruelty ourselves we would have to get ropers who could operate in that incredibly difficult environment of savagely steep slopes and frighteningly high ridges—not only ropers who were highly skilled but also ones who had practiced their skills without, of course, killing but also without even inflicting any harm. And the very first job of all was to find someone who would not only find such people but also actually lead them.

After a long search I was directed to a former World Champion roper by the name of Dave Ericsson. I found him in a small town in Arizona named Wickiup. I had been told before I met him that he was a man who could do anything with a rope, but the one thing he could do which impressed me the most was that he had, I was told, once roped a rabbit. When I faced him with this achievement or rather, from my point of view, with this charge, he was clearly nervous about it because he had obviously been informed about the Fund for Animals' beliefs. At first he tried to get out of it by saying that it wasn't all that big a deal, that the rabbit had been quite near, and anyway he had done it on a bet, and finally that he was lucky. I was immediately impressed with his modesty, but far more so when he also solemnly promised me that he had not hurt the rabbit and that, after roping it, he turned it loose and, as he put it, "Cleveland, it ran off just fine."

In the talks Dave and I afterwards had at his ranch, one

of the things I liked best about him was the way he talked about animals. For one thing, he told me that he wished people would leave wild horses and wild burros alone out there, because he wanted his children to be able to see them when they grew up. For another thing, he told me that he had brought up his children to believe that at his ranch no one should sit down to eat breakfast in the morning—or dinner at night, or for that matter, any meal—without having first been certain that every animal on the place had been fed.

After that talk I knew I had my man, and I told him exactly what we intended to do—to stop the Park people from shooting the burros by offering them the alternative of allowing us to rescue them. For a long time Dave said nothing, and then he asked me what I was paying. This time it was my turn to say nothing for a while. The Fund for Animals had been in existence for only a decade, and we were by no means long on money. But we did have one ace in our hole—the same advertising firm who had helped us get started, and who had coined the phrase "Animals Have Rights, Too." The firm of Young and Rubicam had, I told Dave, come up with what I thought was a wonderful ad which they agreed to run for a nominal price on a whole page in *Parade*. The ad, I explained, would show me holding a baby burro on the rim of the Canyon and the copy would read simply, "If You Turn This Page, This Burro Will Be Shot." Dave was very impressed by the ad and I told him luckily other people had been, too.

The response to the ad enabled us to pay Dave not what he wanted, but at least enough to get him. Dave promised me that he would put together a team that would not only do the job but also do it, he also promised, the right way. He also told me that for the roping he would have to have not only horses, but also mules. When I asked him why he would need mules, he told me because mules were smarter than horses and would only put their hooves down where they knew it was safe. He said they would have to ride along a lot of tough ridges in the Canyon, and unless they knew where to put their hooves down they not only might be hurt, they could also be killed. "Mules always know where to put their hooves down; horses don't," he said. When I asked him why, Dave smiled. "The reason mules are smarter than horses," he said, "is because their fathers were burros." He looked at one of his children, who was doing something he didn't like. "And another reason they are smarter," he added, "is because they don't have children."

One piece of news that galvanized our determination to do the job was that we learned that the Park people had made their decision to shoot the burros based on a report from a wildlife biologist. We were, of course, by then well aware that wildlife biologists in general have one answer to any problem involving wildlife, and that is to shoot it—not the problem, unfortunately, but the wildlife. But what made this report so especially infuriating to us was that we also learned that the man who had written the report, and was therefore the man primarily responsible for the Park people having decided to shoot the burros, was the same man who was going to be paid for running the shooting—indeed, he himself would be one of the ones doing the shooting.

That was really too much. We well knew, to begin with, that the burros are very difficult animals to shoot. For one thing, they are extremely intelligent, and the minute the shooting started they could be counted upon to find every possible hiding place in a terrain far more familiar to them than to their prospective murderers. For another thing, the burro has only two vital areas—the brain and the heart—and the shooters, who would obviously have had little experience with either, would certainly wind up wounding many more burros than they would kill outright. Dave himself told me that he had witnessed one shoot in which he saw a burro, as he put it, "one who had ten bullets in him"—trying to die.

BY THE TIME WE HAD BEEN GRANTED PERMISSION TO attempt our rescue, the relations between us and the Park people had become severely strained. This was not just because of us, either; it was because we had become to them indelibly identified with the burros, and to them anyone or indeed anything that had anything to do with the burros, or particularly anything to do with being in favor of them, was an enemy. A remarkable example of this was a little book written back in 1951 by an author named Marguerite Henry. It was a touching fictional story of a lonely burro in the Canyon named Brighty who befriended, and was in turn befriended by, as the book jacket says, "a grizzled old miner, a big-game hunter, and even President Teddy Roosevelt."

"But," this jacket copy goes on, "when a ruthless claim jumper murdered the prospector, loyal Brighty risked everything to bring the killer to justice." Actually, of course, being befriended by a big-game hunter, and then by President Teddy Roosevelt, who shot everything in sight apparently, with the exception of little Brighty, was a tough enough life to begin with, never mind the murder. But in any case the saga of Brighty for many years had been a local classic in the Grand Canyon area and did a brisk business in the Park people's shops. Indeed, they even put up a statue to Brighty, which someone had given to them, in an extremely prominent place.

Once they had decided to shoot the burros, however, all this came to a crashing halt. No more could you find hide nor hair of the story of little Brighty in any of the shops and, as for the statue of little Brighty, it was nowhere to be seen. Once we had rescued the burros, however, back came little Brighty to the stores again and, sure enough, back up went the statue on a prominent Canyon rim. Even today, long after the rescue, if you should go back to the Canyon and ask the Park people whatever happened to the burros who used to be down in the

Canyon, they will tell you that they were all rescued. And if you go on and ask them who rescued the burros, they will tell you that, too. They will tell you they did.

Not only did the Park people never rescue so much as one burro, they also put every possible obstacle and roadblock in our path that they could think of. Perhaps the most ingenious of these was the "quota" they gave us. Although no previous rescue had ever, in any length of time, gotten out more than twenty burros, our "quota" was fixed at thirty burros within a month—otherwise there would be no more rescue.

On top of that, and as a further way of trying to make sure we did not succeed, they set our starting date for the first day of the rescue on the ninth of August, the hottest time of the year in the Canyon and what turned out to be further bad luck for us, the hottest time of one of the hottest summers in Canyon history. It was so hot indeed that down there on the floor of the Canyon, where the temperature was as high as 120 degrees, even if we had been able to work in the middle of the day we could not have taken a burro up with a helicopter—the heat would not have permitted us the necessary lift. Our only hope was to work in the early morning, as soon as there was at least enough light to see, or late in the afternoon when, again, there was just enough coolness and at least some light. We worked indeed, as Texan Sid Richardson used to put it, "from can to cain't."

As if this was not enough proof of their opposing us, they even put up a further obstacle. We had early found out that they regularly broke promise after promise and agreement after agreement, but this was the worst. It was so bad that it almost made the first day of our rescue also the last. What the Park people had promised us was that we could use, to get our men and horses and mules and equipment down to the bottom of the Canyon, the relatively wide tourist trail to the bottom of the Canyon floor. At the last moment, however, the Park instead gave us a trail that, as I first looked at it that morning, seemed to me to be one that in some places was close to nonexistent and, in other places, straight down. Our men and horses and mules were still within sight of the top of the Canyon when suddenly one horse slipped and fell right into the horse in front of him, who in turn fell right into a third horse in front of that one. Only incredibly quick thinking on the part of Ericsson and one of his best riders saved the day, as well as the animals and our rescue.

Despite the way that first day had started, the second day, our first working day, was memorable. Working, as I said, just early in the morning and late in the afternoon, because of the 120-degree heat, our team managed to rope, tie, put into slings, and helicopter up to the Canyon top no fewer than twenty-seven burros. In one day we had gotten ninety percent of our first month's quota.

Our success began attracting sizable crowds of people, who would gather around our corrals at the top of the Canyon to look at the burros, protected from them by a fence. At once, at the sound of the helicopter, with a burro in a net underneath, a cry would go up, "Burro coming up!" and more people would gather to watch. What they saw was a pilot who would gently and expertly let the burro down on his back in the net. Immediately the corral workers would rush to untie the sling and throw it back up to the helicopter. Then, while the helicopter took off, the corral workers and the vet would approach the burro, the vet would give him an examination, and then the corral workers would untie his legs.

It was a moving experience, and while at first the cowboys poked a lot of fun at the "Bambi lovers," as they delighted in calling us, as time went on they did less and less of this. Indeed, one day at the end of the first week when I had come from another part of the Canyon down to the Canyon floor, I noticed a group of cowboys standing around the helicopter. As I came closer I saw that they had that day, for the first time, captured both a mother and a baby burro. I also learned that a spirited debate was in progress about this—one concerning whether the mother or the baby should be lifted by the helicopter first.

One cowboy was adamantly insisting that it would be better to lift the baby first. The mother would then, he said, at least see what was happening and would be relieved when the helicopter returned for her. Another cowboy was equally adamant that the mother be lifted first. That way, he said, she would at least know she wasn't being hurt, and that maybe her baby wouldn't be either. Finally it remained for Ericsson himself to settle the matter. First he asked the helicopter pilot what was the heaviest male burro we had lifted so far. "One was close to six hundred fifty pounds," the pilot replied. Ericsson next asked the weight of the heaviest female burro he had carried. "About four hundred pounds," came the response. "Okay," Ericsson said. "How much does a baby weigh?" "I'd say a hundred fifty," said the pilot. "Hell," said Ericsson, "let's build two slings and lift them together."

Our "Bambi loving" had apparently spread. In any case, we were told by several veteran rescue authorities up at the corral that we had engineered the first helicopter rescue of large animals ever when the mother and the baby had been lifted together. Altogether the operation took two years and involved saving 577 burros. We had some close calls, but to my knowledge not a single burro, horse, mule, or rider was badly injured. This was due in no small part to the extraordinary skill of the helicopter pilots who, day after day, braved terrifyingly different conditions and dangerous swirls and eddies of winds to land every single one of their precious cargoes lightly and with pinpoint accuracy. The skill of the riders, too, and the courage of their horses and mules, going over the difficult ground, was awesome. As for the ropers, their talent too was superb.

I have never forgotten one example of this. My assistant, Marian Probst, came out to view the rescue and Dave Ericsson rode up to her as she was sitting on the grass down on the floor of the Canyon and told her to stay right where she was. He said he would soon have a burro for her to pat right at her feet. He was showing off, of course, but it turned out to be true. Only a few moments later a burro came running by, trailed by Ericsson and his horse. Then, at the very moment the burro shot by Marian and me, out curled Dave's rope, and the next instant there was the burro, quietly standing only a few feet from our feet. So near, indeed, that all Marian had to do was to stand up to pat him. And then still, just a few moments later, the helicopter appeared, one of the cowboys grabbed the sling, the burro's hooves were tied, he was trussed into the sling, and then the next moment he was sailing off from us, up into the sky.

There were also many humorous moments. One of these occurred at the Indian corral on the Colorado River, an area where the burros did not have to be helicoptered out but were merely roped, placed in barges, and sailed down the river to the corral. But the corral was not on Park land; it was on Indian land. While we generally had much better relations with the Indians than we had with the Park people, they were not in the slightest way shy about getting as much money from the Fund for Animals, charity or no charity, as possible.

They had decided, for example, that since the corral was on their land they would charge us ten dollars for bringing a car down to the corral area and parking it. One day I came to it for the second time that day. The Indian sitting, as usual, by a chain across the gate asked me for ten dollars. I told him I had been there earlier—as I knew he well knew—and I confidently started to go by him. It was no use. It was Indian property, he explained, and it was ten dollars each *time*, not each *day*.

He had me. But I am from Boston and, when it comes to money, Indian or no Indian, I had not given up. I pointed over to the corral where there were many Indians, some of them looking at the burros and others just taking pictures. Who are those people, I asked him, and what are they doing? "They're our people," he said, "and they're taking pictures." Oh, I said. What are they taking pictures of? "The burros," he said. Oh, I said again. And whose burros are those? For the first time, he actually smiled. "I guess they're yours now," he said. I told him indeed they were, and furthermore I wanted to know who had given the Indians at the corral permission to take pictures of our burros for nothing? I started to go on, but I did not need to. He smiled broadly, and from that time on it was no longer ten dollars per trip for me, it was ten dollars per day.

THE REAL HEROES OF THE GRAND CANYON RESCUE were, of course, the burros themselves. Those of us who became intimate with burros in the Canyon rescue for the first time never forgot the affection for these animals which started quickly and then, as you got to know them better, continued to grow. The Fund's second large burro rescue a few years later, in another National Park, the Death Valley National Monument, was far larger in numbers than the Canyon rescue, involving thousands of burros. But even being associated with that rescue and getting to know so many of them did nothing to lessen the individual affection and respect we had for each and every one of them. Indeed it made it, if anything, grow even more.

To many animal activists, the Fund for Animals' attention to burros seemed out of all proportion in terms of cost to how much we could do to help save other animals in trouble. But to this we had one answer—this was that the burro was the beast of burden for the whole world, and that the case could easily and effectively be made that no other animal had ever suffered so much, for so long, and so unfairly as the burro, that the overworking, the overloading, the underfeeding, and the underwatering had been undoubtedly as bad and perhaps even worse throughout the years for burros than for any other animal on Earth. All in all, we reasoned, if we could make a highly publicized statement for that animal—as indeed we felt we had—then it would be well worth it, no matter what the cost. In our opinion, no animal deserved it more.

I had a memorable experience in this regard when, some time after the Canyon rescue, as part of a foreign-policy group on a trip to Morocco, I had occasion to visit the Casbah and came upon a small, spindly-looking burro groaning under the weight of a large Arab. I walked directly in front of them, stopped them, and asked the man if he spoke English. He shook his head, whereupon not caring whether he spoke English or not, I told him he was too fat for the burro and to get off. Obviously he knew what I was saying and equally obviously was not going to get off. So without further ado I pulled the man off. There was quite a to-do about it all, but the last thing I saw was the burro and the man walking. That evening, however, the members of our committee had been invited to one of the dozen or more castles of the King of Morocco. When we arrived there each of us was ushered up to meet the King by the King's aide. And as it came my turn I was amazed when the King said, "I understand we had the pleasure of your company this afternoon in our Casbah." I did indeed visit the place, I told him, at first rather nervously saying my words, but then I also managed to add sternly that I was sure the burro had found the experience pleasurable also.

The more we explored the subject, the more we learned that no animal throughout his long history has been less publicized than the burro. Indeed, whereas in almost any library you will find hundreds if not thousands of books about dogs and cats and horses, and indeed at least hundreds of books about almost any animal you can think of, you will search in vain for even one history of the burro. After an incredible search I did finally find one burro history, but even that was long out of print. Published by the University of Oklahoma Press, the book is *The Burro*, by Frank Brookshier. Brookshier alone, of any animal author I have ever read, makes clear that the burro alone, of all animals, has been an important figure in three of the great monotheistic religions—Christianity, Judaism, and Islam. The Bible, for example, Brookshier points out, has one hundred and fifty-three references to the burro—no other animal is referred to anywhere nearly as often. Under Mosaic law both the ass and the ox were to rest on the seventh day, and Jews also did not eat the flesh of the burro—in fact, they regarded it as unclean. The burro even made it into the Ten Commandments—albeit as an ass—in the Tenth Commandment about not coveting, coming right after thy neighbor's wife, his manservant, his maidservant, and his ox.

Brookshier also points out that only twice in the Bible does an animal talk—the serpent in the Garden of Eden, and Balaam's burro, who defended herself from Balaam's beating by saying she had seen an angel of the Lord. Later, Balaam's burro was groomed to be the animal who would carry the Messiah, a prophecy that came true to the extent that there was a burro in the stable when Christ was born, that Mary later rode a burro into Egypt, and finally that Christ rode a burro into Jerusalem.

To this day, you can find people who will tell you that the reason for the distinctive cross-markings on the backs of all burros is the animal's historic connection with Jesus. The fact, however, is that burros were present thousands of years before Christ, although it is typical of the burro's anonymity that there has been no accurate census of how many burros there were in the old days or even how many there are in the world today. One thing is certain, though—they exist almost everywhere there are people. Whereas the number of most animals has in general decreased due to mechanization and industrialization, the number of burros has not only not decreased, it has actually increased.

In our own case, during our Canyon rescue as the number of our burros grew we realized how lucky we had been to have bought the Ranch and therefore have a place to put them. At the same time, however, the growing number of burros made us realize that we could not possibly keep them all on our Ranch, even for a short period of time. We would have to have a countrywide program to adopt them, and for this we would need way stations—both as places to give the animals veterinary care and as places from which to do the adoptions.

Wild burros make wonderful pets. That's what we had said before we even began our rescue and well before we had the slightest idea whether they would make pets at all, let alone wonderful ones. But as we began to have more experience with the burros in our Canyon corrals we knew that we would in time, if not right away, be right about them making wonderful pets. First, though, we had to convince our way station people, the people who would have to oversee the adoptions and also convince the adopters themselves of this fact. Fortunately we chose truly extraordinary animal-loving people to oversee the adoptions. Even years after they did their wonderful work some of these people have stayed indelibly imprinted on my mind. Cynthia Branigan, for example, from Pennsylvania; Vicki Claman, from Connecticut; Mary Truland, from Maryland; Nancy Nelson, from Minnesota; and Tom and Judy Currin, from North Carolina.

Not a single one of these people or any of the many others who helped had had, up to the time our trucks with the burros on them arrived, one iota of experience with any burro before. Yet here they were suddenly faced, not only with "wild" burros, but also with burros that had had just the most minimal experience with people. And even those people with whom the burros had had this experience were people who, in their eyes, had to be totally around the bend—people who first chased them on horseback or muleback, then threw a rope around their neck, then tied their legs together, then put them in a sling, and finally, under a frighteningly noisy machine, took them higher up into the sky than any self-respecting animal and perhaps any reasonably wise bird had ever been before.

In spite of all this all our adopters agreed that the burros did indeed make as wonderful pets as we said they would. The adjectives these people used to describe their charges were "intelligent," "curious," "philosophical," "benevolent," "playful," "thoughtful," and, one that especially appealed to me: "humorous."

In the contract we had with our adopters we insisted that not only must they take two burros—or one if the adopter had a horse or a goat or something so that the lone burro would not be lonely—but we also insisted both that they write us regularly and also send us pictures of their burros. Besides this, we asked that each adopter admit a Fund for Animals person at any time to look over their burros and make the determination that if in that Fund person's subjective view the burros were unhappy we should be allowed to take them back. Our peerless lawyer, Edward Walsh, said that the words I had written in that contract simply would not stand up in court, but I instructed him that he was to put them in anyway, for the good of my soul. Besides all this we asked one and all for regular and definite comments on their burros. Some of these would make a book in themselves. My favorite was a simple statement from a young woman: "My burro," she said, "makes a much better pet than my husband." Another letter was in the same vein, only from a married man: "In my next incarnation," he wrote me, "I should like to come back as a burro being looked after by my wife."

Not the least interesting assessment of the task of the adopters was that not a single one of them mentioned a single kicking, butting, bumping, or any kind of harm done to them or any member of their family on the part of a burro. Considering the fact that hundreds of children were parts of the families to whom the burros went and that these children, particularly the young ones, surely climbed up on their burros from the back or did heaven knows what to them and not one of these was ever bothered by the burro surely speaks eloquently for the burros' patience, fortitude, and truly remarkable understanding of children. All in all there was only one criticism of burro behavior in our whole experience and this was the fact that the burro did indeed kick, butt, bump, and do just about anything he could except actually bite when he was being put in a truck or a trailer for a trip. Here the burro proved himself a tough customer indeed—the problem being that for one thing he or she did not know where he or she was going, and for another he or she was not at all sure he or she would like it as well as where he or she was, and for a third thing he or she would like to wait and see, and no one seemed to give him or her a chance or even an opportunity to make up his or her own mind.

One virtue of the burros that especially endeared them to our adopters—at least the ones who had other animals on their farms or ranches—was the fact that the burros would guard other animals from predators—even animals of whom the burros were not particularly enamored. Furthermore, the adopters told us, this kind of guard duty was performed not just by the male burros but also by the females. As for a female protecting her baby, a remarkable number of our adopters were familiar with stories of female burros fighting off animals even as large as mountain lions to protect their young.

Since many of our adopters were familiar with horses but had not previously been familiar with burros, they were particularly interested in the differences between horses and burros. More than one said they would never use the phrase "horse sense" again, and certainly not in the presence of a burro. The phrase they said they would insist upon from then on was "burro sense." Their examples of the differences between horses and burros were legion. Burros, they found out, would never be guilty of overexertion or overdrinking or overeating. They learned, too, that burros do not like to be led the way horses do, and especially do not like to be led, or rather pushed—as they must be in this case—backward.

A curious difference almost everyone who had to do with the burros found out about sooner or later is that between the neigh or whinny of the horse and the burro's bray. The neigh or whinny of the horse is interesting enough, but the bray of the burro is truly never to be forgotten. Jonathan Swift called the burro "the nightingale

KEITH MCELVAIN

Entrance to Ranch. The sign at left reads: "I have nothing to fear, and here my story ends. My troubles are over; and I am at home. *Last lines of* Black Beauty, *Anna Sewell.*"

MARY DA ROS

Peg, the first customer of Black Beauty Ranch, stands here at the monument to Polar Bear. The monument reads: "Beneath these stones lie the mortal remains of The Cat Who Came for Christmas, Beloved Polar Bear. 'Til we meet again."

THE FUND FOR ANIMALS

(Left) The Fund for Animals' most famous rescue—of 577 burros in the Grand Canyon—all of whom the Park Service wanted to shoot.

The author and one of his favorite subjects.

CRAIG FILIPACCHI

MARY DA ROS

Two of our many beautiful barn cats—all spayed and neutered, of course.

MARY DA ROS

These baby raccoons, after being rescued, were taken to the Ranch's 250-acre wildlife area.

MARY DA ROS

Some of the goats rescued from the U.S. Navy's San Clemente Island here enjoy their age-old game of King of the Castle.

The Fund for Animals

140 West 57th Street, New York, New York 10019 VOL. 15, NO. 4

At Work . . . at Home, Abroad

FUND SAVES LAST DIVING HORSES

"PRICE IS RIGHT"-- NO MORE FURS!

"PAW IN THE DOOR"-- FOR SENIOR CITIZENS

NEW "MONKEY TRIAL" ROCKS LABORATORY ESTABLISHMENT

THE FUND FOR ANIMALS

A stupid Atlantic City act the Fund for Animals was able to stop. We rescued two of the last diving horses, but this one, Powderface, was sold to a slaughterhouse before we could save him.

A peaceful day at the Ranch of Dreams.

MARY DA ROS

Cody, rescued from outrageous cruelty, and his friend and protector, the late great police horse, Stanley.

MARY DA ROS

Chieftain, chief of Black Beauty Ranch's mustangs.

CHRIS BYRNE

Nim and the late great Sally.

MARY DA ROS

KEITH MCELVAIN

Nim and his new friends—from left, Midge, Kitty, LuLu Belle, and Nim—enjoy a snack between play.

of the animals," and while this has some hyperbole in it, almost anyone who has been with a burro for some time realizes the burro has three distinct brays—one to express happiness or just plain contentment, another to express sadness or perhaps loneliness, and still a third for hunger. Besides these there is of course the very special bray of a mother for her baby, and the equally special one of a baby trying to learn to bray.

I became lifelong friends with many of our adopters and still correspond with several of them. Two special favorites of mine are Vicki and Allyn Claman, of Connecticut. "When my husband first suggested we become an adoption center for the Fund for Animals' burro rescue," Vicki wrote me, "I thought our early retirement had affected his brain." Before she began, she also wrote, the only burros that she had ever seen before she took on the job were on the big screen in "B" movies. "I was just wondering how we would be able to handle several of them at once—I imagined, say, three—but all of a sudden eighty-three confused, thirsty, hungry burros arrived on a large two-decker animal transport truck." From the moment Vicki and Allyn and one adolescent boy with, as they put it, "strong but gentle hands" went to work and looked after more than 450 burros, almost all of whom—except for a few they could not bear to part with—ended in adoption homes all over the Northeast, not a single person or burro was hurt, as they also put it, "even slightly."

Mary Truland, our peerless adopter in Maryland, had only one problem with her position. "I find myself," she told me, "more and more reluctant to put them up for adoption at all. Day by day I get more and more fond of each of them." She told me that she had a standing agreement with every single one of her adopters that if they ever called her up and wanted her to take a burro back, she would come to pick him or her up with no questions asked. She also noted that while this situation was rare, when it did happen it was very interesting. "When we brought one back," she said, "he or she was greeted by the others as if they had done a stint in the military or been off at boarding school. In any case, right away they were integrated back into the herd with much fanfare, folderol, and storytelling."

Mary Truland had an additional postscript of which I was especially fond: "People come over and ask me what you can *do* with a burro and I always say, 'Not much—

the same as with cats and, of course, dogs.' I also say, 'Don't kid yourself if you think you will have this burro driving a children's cart in four weeks. You won't. Even if you teach a burro to pull, the burro would have to do it all the time not to revert. But—if you want a good friend who will be a companion to your family and your livestock, and guard your fields and your gardens, as well as warn you when anyone comes up the drive, a burro is for you."

Tom and Judy Currin, in North Carolina, were two other particularly dedicated adopters who themselves had one burro they knew from the moment the shipment came to them they could never adopt "out." His name was Gus, and he came from the Death Valley rescue. He had been shot by the Park Service in the Valley, the bullet having gone into his cheek, through his nose, and into his throat. The Currins had Gus for ten years until he died, and they still miss him every day. Gus had terrible trouble eating, and often food would come out through his nose. And when he drank water, he had to drink by throwing his head back.

Despite all his disabilities, Gus had what the Currins described as "unforgettable, innate dignity," and such proud independence that he would go where he wanted to go, even if it was not, as both Currins put it, "exactly where you wanted him to go." But there was one thing at which Gus was perfect, the Currins claimed—and that was as a baby-sitter. "He was the best we ever had," they said. "When our little Olivia was just a tiny tot and was out in her baby carriage, Gus would hurry right over and curl up right beside her carriage, just like a dog, and woe betide any stray dog or stranger of any kind who came anywhere near her."

Long after Gus was gone, the Park Service in Death Valley was still determined to shoot burros. But thanks to Gene and Diana Chontos, from Onalaska, Washington, who feel about burros exactly as the Fund for Animals' adopters feel, they have been dissuaded from such killing. Instead, the Chontoses made a deal with the Park Service that if they would stop shooting burros the Chontoses would rescue and adopt any number they said they were going to shoot.

Many of the burros we rescued were for one reason or another brought to Black Beauty Ranch and kept there. A whole herd from the center of the Canyon called the Shiminu herd we kept simply because they were so distinctive, with their dark brown coats and bright white circles around their eyes. Another reason was from the beginning the burros were our favorite animals at the Ranch, and we could not bear the idea of parting with all of them. At first this may well have been because they were, except for Peg the cat, our first animals at the Ranch, but as time went on we found ourselves not only respecting them—and their fascinating, philosophical ways—but very soon also developing real affection for them and, finally, actually loving them. Curiously the same progression of feelings toward burros seems to work with a majority of the people who come to see the Ranch in groups and on tours. At first when we tell them "Now we're going down to see the burros," the best we usually get is something between yawns and looks of disappointment. But when they see the tree-shaded low water crossing where the burros like to gather, or just see them spread out over their huge pasture, and then get to pat and hug them, they too just like us become first respectful, then affectionate, and finally begin to love them just the way we do.

As for ourselves, although from the beginning we had many favorites among the burros, we had one *favorite*

favorite. She became this because she was the first at the very early stage of the rescue not to get as far away from us as possible in the corral but actually, first, to stand where she was and hold her ground, second, not to move at all when people came even closer to her, and third, to come closer herself.

I named her Friendly, and one evening when I went to find her in the corral to show people how correct her name was, although I kept walking up to burro after burro in the corral I could not find her. As I looked I heard the cowboys sitting on the corral rail laughing at me, but as they often laughed at me, at first I thought nothing of it. But when finally their laughing veered on the uproarious side I swung around to stare at them and there was Friendly—the whole time she had been plodding along behind me, for all the world and particularly to all those cowboys looking for me as hard as I had been looking for her.

Friendly had come up in a sling under the helicopter in the very first batch of burros we rescued in the Grand Canyon. I was in the corral when she was lifted up over the rim and delicately dropped to the ground. I was also one of the crew who untied her. Whatever had happened to her—the roundup, being tied in a sling, being picked up and fastened to a roaring helicopter overhead, being carried seven thousand feet up in the air, and then finally being let down among all of us and hundreds of onlookers—had all been sudden and uncomfortable and ridiculous and even crazy, and she must surely have thought that was what we all were. But she also realized, I felt then and still feel, that no one had really hurt her, and therefore we were not all bad.

We had already publicly said that wild burros would make good pets—better pets, indeed, we had claimed, from the point of view of temperament, than horses—and we had also said that, rather than get your kid a pony, get him or her a burro. We said that because we needed to get homes for our burros, but at the same time we didn't really know just how good they would be, and we certainly hadn't had it proved to us. The burro who proved it was Friendly. That was why, then and there, in those first moments at the corral when she stood and looked at us and had not trotted away, I had given her her name. In a way I felt Friendly was responsible not only for the successful adoption program but also for the much larger later rescues at the Naval Weapons Center at

China Lake and Death Valley—more than five thousand burros in all.

Friendly was one of the first burros we brought to the Ranch. When we brought her there, however, we soon discovered she was pregnant, and in good time—burros have an even longer gestation period than horses—she gave birth to a burro we called Friendly Two, although she preferred Too. The first time I ever saw her baby was when I was visiting the Ranch and, as usual, had gone over to see her and at that very moment she spotted me and trotted over with her baby from a distant pasture. Almost always when I see Friendly she has a regular greeting—of pushing her head into my stomach. This time she had started to put her head there all right, but suddenly she stopped, moved back, and instead—with some pride—pushed her baby toward me. Immediately I started hugging and hugging and gushing and gushing over the baby until suddenly, even more quickly than she had stopped before, she pushed her head in again, pushed her baby away, and pushed her own head hard back into my stomach. It was just as if she was saying, in no uncertain fashion, that she wanted to show me her baby all right and even wanted me to hug her, but enough was enough. I was not to forget that she was Friendly One, and her baby Friendly Two, and she would thank me very much to remember not to forget that.

To this day Friendly One, and sometimes Friendly Two, will, when I am visiting the Ranch, clump over to the veranda at the main house at four o'clock for tea, tidbits, and gossip.

OUR SECOND ANIMAL WAR WAS NOT WITH THE National Park Service but rather with, of all adversaries, the United States Navy. And this war broke out not over burros but, of all animals, the Navy's own mascot, the goat.

The goats were San Clemente goats—not Richard Nixon's inland San Clemente but San Clemente Island. They were a rare breed of Spanish Andalusians whose problems began in earnest when, during World War II, President Franklin Roosevelt, in a burst of typical pro-Navyism, gave San Clemente Island outright, lock, stock, and barrel, and apparently in perpetuity, to the U.S. Navy.

The Navy had immediately put the island to use as a

shelling target not only from ships at sea but also for field trials of new weapons from the land and the air—something they continued to do after the War. How the goats had been able to survive four decades of this shelling was a mystery and the terrain of the island, with its rugged and craggy protective cliffs, only partially explains it. More important was the goats' uncanny ability to find and take immediate cover when the shelling began. This prowess, we would soon learn firsthand, matched that of experienced troops in battle.

The Fund learned of the Navy's idea of shooting the goats in a small item in the inland San Clemente newspaper. What the Navy proposed was using, not even sharpshooters as the Park Service had at least done with the burros, but just ordinary hunters who would be brought to the island and have some wonderful sport. In any case we started with a lawsuit as we had with the Park Service and this time, in contrast to the burro war, we won. In all, during the Fund's rescue of more than four thousand goats, we were at war with the U.S. Navy in court and on the island for more than six years.

To me, looking back at it, there were three particularly memorable incidents. The first began in the halls of the

Pentagon itself. Here, with Dana Cole, our California lawyer, I made my way toward the office of Vice Admiral Tom Hughes, Chief of Naval Logistics and apparently also Admiral of the Goats. Among the offices we passed was that of Secretary of Defense Caspar Weinberger. He was a man who had preceded me by a couple of years as president of the *Harvard Crimson* and also one whom, through his assistant Benjamin Welles, we had long been trying to get to intercede for us in favor of the goats. As we walked we could not fail to see that along the whole incredible corridor was oil painting after oil painting apparently depicting famous sea battles. They pictured one ship after another blowing up other ships. All of those, of course, doing the blowing up were our ships, and the blown-up ones were enemy ships.

Finally we turned into the admiral's conference room. Here there was a full complement of captains, lieutenants, ensigns, and civilian personnel. Shortly thereafter Admiral Hughes appeared, introduced himself, shook hands, and sat down. First he lit his pipe and then, without preamble, rose and went to the large map of San Clemente Island on the wall. From here things began to go from bad to worse, especially when the admiral brought up the question of the endangered flora and fauna on the island. Whenever the admiral said anything, all the others at the table nodded and said, almost in unison, "Aye, aye." Finally the admiral went over and sat down. "You see, Mr. Amory," he said, "our hands are tied. We have to get rid of the goats. They're eating endangered species. It's the Endangered Species *Law.*"

I had been good so far, but bringing up the Endangered Species Act was too much. No one had fought harder for this Act than the Fund for Animals, and for it to be used now against us, and on such flimsy grounds, was truly infuriating. I pointed out that, as I understood it, there were exactly three endangered species of fauna on the island and that, as I also understood it, two were birds and one was a lizard. I told the admiral that I was prepared to admit that an occasional goat might occasionally step on an occasional lizard, but I would certainly like him to name an instance of a goat eating a bird. I also asked him if his gunners, when they shelled the island, took care to avoid both the birds and the lizard.

There was a long silence. Finally I could stand it no longer. I looked first at the admiral and then at the others. Dana remembers that I called them a bunch of murderers, but I think I phrased it much better than that. I did, however, mention for the benefit of the Annapolis men present that what they were shooting was their own mascot. In any case, the result was predictable. Dana asked me to leave the meeting.

After the meeting, Dana suggested that, having failed with the Navy brass, we should now take our case to the media—which had been consistently favorable to us and the goats. We started with the ABC network. On the way I told Dana what I proposed to say—that I now knew why the Navy had botched the rescue of our hostages in Iran. "How?" Dana asked. It was because, I said, they evidently thought they were going over to shoot them—they didn't know they would have to rescue them.

"If you say that on the air," Dana said to me slowly, "you will have to get yourself another lawyer." Don't worry, I told him. I would never do such a thing. And today, as I look back, I really think I intended not to say it. But, when we reached the network, all the effort we had put into the goats somehow got the better—or worst—of me. "What do you think of the Navy turning you down?" the interviewer asked. I know now, I said firmly, why the Navy had botched the rescue of our hostages. They thought they were going over to shoot them—they didn't know they would have to rescue them.

"That's great," the reporter said enthusiastically, and immediately rushed out of the room. In a few moments he was back. "We're going to lead the news with it," he said excitedly, and with that he once more went away. I looked at Dana. This time he refused even to look at me. It was, however, almost time for the evening news and for some moments we just sat there waiting to see it. Suddenly, just before airtime, the reporter appeared again. "Here," he said, "look at this." He showed us a news clip. The clip said that Secretary Weinberger had overruled his admirals about the goats and that the Fund for Animals would be allowed to rescue them after all. "Now we're going to lead with this," the reporter said, almost as excitedly as he had before. At first I too was so excited that I couldn't think of anything but our victory. But then, suddenly, I remembered what I had previously said about the Navy. I told the reporter I hoped he would not use it. "You'd better," Dana said, "hope hard."

Mercifully, they did not use what I had said. In any case, the second most memorable incident—the rescue itself—was shaping up in California. Our first job was to figure out how to take care of all the media who wanted to be in helicopters to take pictures of the country's first rescue with both helicopter and the then brand-new "netgun," as it was called. What this was was a gun that fired, not a bullet, but a net blanket over the animal—one that had weights on all four corners of the blanket and would, if properly aimed, drop over the animal to stop his struggles. In any case, what Mel Cain, our peerless pilot, and New Zealanders Bill Hales and Graham Jacobs, inventors of the netgun, did that day on that island—with a dozen news helicopters buzzing around them—was truly extraordinary.

Mel always insisted on either an up or level run, not a down one in which a goat captured in the net could hurt himself—and the first goat was beautifully and accurately run uphill. While Bill jumped to the ground to tie the goat's legs, remove the net, and put the animal in the sling, Mel made a quick circle. Then, when he returned, Bill quickly hooked the sling to a ring under the helicopter, and the goat was then flown back to the corral where he was gently landed and untied. The whole operation had taken less than four minutes. In all that day more than sixty goats were rescued, some from very difficult terrain.

The third and final memorable incident of the rescue occurred when, later, we were allowed to enter the hitherto totally forbidden unexploded-shell area. Here I was alone with Donna Gregory, the Fund's corral boss, one morning down at the very end of the island. We had a truck there that we were using as a holding corral and were awaiting Mel and Bill's first morning run. All of a sudden, bouncing along the rough dirt road toward us we saw a Navy jeep with two naval officers sitting in front. As they turned up the hill toward us at full speed, I didn't know what we had done wrong, but I told Donna it sure looked as if we would shortly be headed for the brig.

Finally the officers arrived and jumped out of the jeep. "Mr. Amory," the senior officer said excitedly. I "aye aye'd," "sir'd," and all but saluted him. "Mr. Amory," the officer went on, "have you or Miss Gregory seen any suspicious-looking ships in this area?" Ships? I automatically queried. "Yes," he said, "we've had reports of a Russian ship in this vicinity, and we wondered if you'd seen anything like that." No, I said, quickly adding, "No, sir." "All right," he said, "but keep your eyes open. And if you see anything at all, take your truck and drive right to the command post." With that, both officers jumped back into their jeep and took off without another word.

It was all Donna and I could do to keep from laughing before they were out of earshot. Here we were, on a U.S. Navy–owned island, far out in the Pacific—an island bristling with every kind of radar and detection device known to modern warfare. And yet our entire national security apparently depended upon whether or not Donna, who wears thick glasses, and I, who on a good day might be able to see fifty yards, saw a Russian ship. It gave one pause, all right—in fact, as I looked at Donna, I could see it gave us both pause.

One enduring dividend that came to us from the goat rescue was a whole new California shelter. In looking around for a place near San Diego from which to do our goat adoptions, we happened upon the Animal Trust Sanctuary in Ramona and its formidable founder, Patricia Nelson. Ms. Nelson not only insisted that our goats be adopted from her shelter but also, upon her retirement, gave it to us—lock, stock, and barrel. To run this, we installed an eighteen-year Navy civilian executive named Chuck Traisi, a man who from the beginning was so outspoken against the Navy's shooting the goats that he first joined us in taking part in the rescue and then, refusing to wait two years for his pension from the Navy, joined the Fund itself. With his wife, Cindy, he now not only directs what has become one of the country's finest wildlife rehabilitation centers but also operates something that is particularly dear to my heart—a cattery that even Polar Bear would have found difficult to criticize.

Just as we had done with the burros, as soon as our rescue of the goats was over we set about arranging to have them adopted. The only real change we made was that whereas with burros we allowed one to a customer if they had horses or some other animals, with goats we demanded that people adopt at least two goats. Goats are herd animals, and we wanted them to have at least one other of their own kind with them. When it came to the promotion of our adoption program, we also went about it with the goats just the way we had with the burros.

Although I was never as intimate with the goats as I was with the burros, I do remember one television program in Los Angeles where I launched into a veritable paean of praise about wild goats as pets. I even went so far as to forget about my beloved burros entirely and made the statement that wild goats were the best pets there were. At this the hostess of the program decided I had gone far enough. "Mr. Amory," she said sternly, "the last time you were on this program you said wild burros were the best pets there were. This time you say wild goats are. Which is it, Mr. Amory, one or the other, burros or goats?"

She had me, but I had to say something, and I also tried to be as stern with her as she had been with me. "Whichever," I said, "we have the most of."

Chapter Three

OF HORSES AND THEIR HISTORY—WILD, DIVING, DOMESTIC, AND OTHERWISE

A RANCH BEARING THE NAME BLACK Beauty could hardly be without horses. Our Ranch may not have started out to have been, first and foremost, a place for horses—that position was reserved, as befitting our most famous rescue, for burros. Nonetheless, from the beginning, our horses were, in the minds and hearts of all of us who had developed the Ranch, second only to our burros and came indeed not only to outnumber them but also to outnumber any other animal on the Ranch. From the beginning, too, we either attracted or rescued a truly extraordinary variety of horses—thoroughbreds and plough horses, racehorses and show horses, Arabians and Appaloosas, Morgans and miniatures, ponies and paints.

They came in every conceivable color—blacks and whites, grays and bays, chestnuts and pintos, sorrels and strawberry roans. One thing, however, almost all of them had in common—by the time they came to us they were, as were the other animals, either old and infirm, lost or abandoned, abused or misused. And every single one of them would learn, as the horse Black Beauty had put it in the last lines of the book, that, when they came to the Black Beauty Ranch, they had found a home which would be theirs for the rest of their lives.

The history of the horse in North America is a fascinating one. In the first place, all modern horses are believed to be linear descendants of the original horse, *Eohippus,* whose bones have been found in both Great Britain and America. In the second place, although the bones of *Eohippus* existed almost fifty million years ago, they were fairly recently discovered in, of all places, the Texas Panhandle. In the third place, although *Eohippus* was only ten hands tall—too small to defend himself and not fast enough to escape his enemies—he nonetheless managed, somehow or other, to exist for fifty million years. In the fourth and final place, after existing for all those years all of a sudden, like the dinosaurs, the horses disappeared. Furthermore, they disappeared along with the very companions with which they had lived—animals that ranged from dinosaurs to camels.

As in the case of the dinosaurs, historians have never been able to come up with a reason for this disappearance—with the single exception that, generally speaking, they believe Man was the culprit. Whether *Eohippus* and his ill-fated companions were hunted to death, raced to death, or eaten to death is not known, but the probability is that all three of these possibilities played a large part in their demises. In any case, what is known is that all of a sudden, and as suddenly as the horse disappeared, he and his companions reappeared—8,000 years after their disappearance.

Furthermore, the horse reappeared, at least in America, under the aegis of no less a renowned figure than the explorer Hernando Cortés. In 1519, on his first voyage to Mexico, Cortés brought horses with him, if for no other reason because, as he himself put it, he valued the life of a horse more than the lives of twenty of his men. When Cortés left Mexico for Seville he took what horses he had left with him—all except the foal of one mare who escaped and went wild. In 8,000 years she may conceivably have been, in all senses of the word—with the presumed aid of the stallion of some follower of Cortés—the founding mother of America's first wild horse dynasty.

Unfortunately the wild horses, or mustangs as they had come to be called, would undoubtedly have been better off if they had never reappeared, at least in North America. Although in time they grew far taller than their ancestors had been in the latters' previous existence, and although they were at least at first regarded as a vital and interesting phenomenon of the Old West, that feeling was all too short-lived. Indeed as the nineteenth century went on, and the cattlemen and sheepmen became more and more powerful, they decided that the wild horses were predators, if not of the cattle and sheep themselves, then at least of their grazings. In time these cattlemen and sheepmen used every means possible to exterminate the wild horses, even though, ironically, one of the main reasons for the increasing numbers of these horses was that in bad times, the cattlemen and sheepmen often simply released their own horses to fend for themselves and these, joining up with wild horse bands, became wild themselves.

No method was too cruel for the ranchers to use in their zeal to exterminate the wild horses—from poisoning their watering holes to blinding lead stallions by shooting their eyes out or just by running them to death, off and over cliffs. Not the least cruel method of wild horse capture was taking a group of already captured mustangs, sewing their nostrils together with rawhide so that they could barely breathe, and then returning them to another band. Here, since they could only run slowly, particularly in the heat, they would slow down the entire new band and make the new band's capture that much easier. As if all this was not enough, finally, in 1897, the Nevada legislature passed a law allowing wild horses to be shot on sight by any citizen.

The twentieth century was no better for the wild horse than the nineteenth. The cruelties were still inflicted on them, perhaps most memorably in the famous movie *The Misfits.* Ironically this movie, Clark Gable's last as well as that of Marilyn Monroe, befitted its title all too well since both stars, at least when they were working together, were so obviously miscast. Nonetheless, the portrayal of the mustangers and their brutality was unforgettable—except, of course, to the western ranchers, who dismissed it as being exaggerated and of course did nothing to address the problem.

ONE NEVADA HOUSEWIFE, HOWEVER, WHO HAD MARried a rancher, did do something about it. Her name was Velma Johnston. Small in stature, weighing less than a hundred pounds, she was a woman who had as a child suffered a severe case of polio which left much of her body and even part of her face misshapen. Nonetheless she also had had from childhood a remarkable combination of persistence and courage, and these traits were combined with a blazingly deep hatred of cruelty, all of which made her, from the moment she saw her first wild horse, one of the greatest animal friends ever.

When I first met her, she told me that although she had lived almost all her life in Nevada, where there are more wild horses than in any other state, it was relatively late in her life when she first saw one. "The fact that I never saw one," she said, her eyes sparkling with irony, "shows how much trouble to the ranchers they really were." Nonetheless, from the day she saw her first wild horse she never gave up her fight for them—even though, ironically, she saw that first wild horse purely by chance. What had happened was she was driving one day when she noticed in front of her car a truck carrying what she assumed to be a load of animals of some kind. But what she also noticed about this truck was that it was leaving a trail of blood.

That was enough for Mrs. Johnston. She followed the truck and, when it finally stopped, got out and demanded from the driver that she be allowed to look inside. The truck was literally crammed with what the driver told her, in a relatively unconcerned fashion, were wild horses. To her dying day Velma Johnston never forgot what sight greeted her eyes in that truck. One little foal had actually been trampled to death in the crowding of the animals. Another had no hooves at all. Where the hooves should have been had been worn down to bloody stumps. A stallion stood motionless, looking toward her, except that his eyes were shot out—he was totally blind. When she had seen all she could stand, Mrs. Johnston turned to the driver and asked him what had happened to the horses. He told her they had been "run down," as he put it, by airplanes.

Mrs. Johnston's next stop was at the district office of

the Bureau of Land Management, the agency supposed to "look after" the wild horses and also the office from which permits were issued to "run" them, as it was called, from planes. All too typically Mrs. Johnston would later learn the Bureau director was very pleased to see her. He was also certain that she had come to praise his work. "By letting these profiteers run the horses," he told her, "we've been able to get them off the ranges to make more land available for your cattle and sheep." Mrs. Johnston could hardly believe his words, but the director went on. "It's all done without any expense to the taxpayer," he concluded proudly. "You see," he said, "everybody profits."

Mrs. Johnston didn't answer the man, but she had a special look for people like that. She also told me that she never forgot what that director said. Later, when she told her father what she was going to do, and how determined she was to fight for the wild horses, he did not try to stop her but asked only that she remember four things. "Dress like a lady," he said, "act like a lady, talk like a lady, but think like a man."

Such advice may seem not only old-fashioned but on the Neanderthal side today, but the fact is for that time, in that state, in that particular fight, it was good advice. And Mrs. Johnston abided by it, even when her fellow ranchers refused to call her by her real name but instead mocked her with the nickname they made up for her—"Wild Horse Annie." Tirelessly, with facts and figures and pictures and statistics about the cruelty, she went from legislator to legislator to ask that something be done about the horses. At first she had only one supporter, a state legislator named James Slattery. But on a memorable day in 1955 a so-called pro–wild horse bill, dubbed the "Wild Horse Annie Law," was adopted by the legislature.

Basically what the bill provided was the banning of the use of aircraft to round up or hunt wild horses, and also the prohibiting of the poisoning or pollution of watering holes. Remarkably, the ranchers and Annie's other opponents hardly bothered to fight her law. Their reason was a simple one. In their opinion, and accurately, Annie had spent four years fighting for virtually nothing. The law affected just state-owned land, and since 86 percent of Nevada was Federal land, what it amounted to was that Annie had been able to stop the cruelty in just 14 percent of the state.

By ignoring her law, however, the ranchers had underestimated Annie. She knew full well that her law affected only 14 percent of Nevada, and she also knew full well just how she would make use of that fact. Hardly was the ink dry on Nevada's "Wild Horse Annie Law" when Wild Horse Annie, who now proudly called herself by that name only, was on her way to Washington, D.C., fighting for a Federal law. "Look," she told every Senator and Congressman who would listen, "the poor people of Nevada are trying to protect their wild horses even though they know they can only protect them on just 14 percent of their land. Now they are begging you to allow them to protect those horses on the other 86 percent of their land."

This was actually, of course, some distance from the way all too many people in Nevada felt about wild horses, but Annie was at her persistent best in seeing that the people in Washington got the message anyway. Soon she was proving herself as adept at handling Federal legislators as she had been handling those at the state level. Again she had the facts and the figures, the pictures and the statistics. Besides this, she took advantage of every possibility that came her way.

The most remarkable of these was when a class of nine-year-old schoolchildren from Roseburg, Oregon, under the direction of their teacher, Ms. Joan Bolsinger, wrote to Annie in Washington, D.C., on behalf of the wild horses. Annie saw this as a very special opportunity. Soon she had children in other states also writing. First there was a trickle, then a river, finally a flood—actually, the largest amount of mail from children in Congressional history up to that time.

But Annie did not stop with just letters. She undertook to see that as many children as possible actually followed their letters and went to Washington, D.C., in person, with their mothers with them. She also saw to it that as many children and their families as possible not only personally visited the offices of Senators and Congressmen but also visited them in their homes, where they could personally talk with the legislators' children and get them also to plead for the wild horses. Since it is a bold legislator who will vote against his or her children's wishes, not long after the children descended on the Capitol, Annie had her Federal law. It was called the Wild and Free-Roaming Horses and Burros Act:

"Be it enacted by the Senate and House of Representatives of the

United States of America in Congress assembled," the bill declared in ringing language, "That it is the sense of the Congress that free-roaming horses and burros are living symbols of the historic and pioneer spirit of the West, that they contribute to the diversity of life-forms within the Nation, and they enrich the lives of the American people, and that all free-roaming horses and burros shall be protected from capture, branding, harassment, or death. . . ." There was even an amendment that included, also in two ringing prohibitions, the following:

> *(a) Whoever uses an aircraft or a motor vehicle to hunt, for the purpose of capturing or killing, any wild unbranded horse, mare, colt, or burro running at large on any of the public land or ranges shall be fined not more than $500, or imprisoned not more than six months, or both.*
>
> *(b) Whoever pollutes or causes the pollution of any watering hole on any of the public land or ranges for the purpose of trapping, killing, wounding, or maiming any of the animals referred to in subsection (a) of this section shall be fined not more than $500, or imprisoned not more than six months, or both.*

Sadly, even with all Annie's work the wild horses and burros were not yet really protected. The law was basically a "no-chasing" law, and even that was flawed. The mustangs could still be chased to private land where they were not protected, and no one could prove where the chase had started. They obviously needed much stronger protection, but such was the hold of the ranchers and wildlife management people that more protection looked impossible to achieve.

Indeed, up until the "Wild Horse Annie" law the Federal Government did not even have a category for wild horses and burros. The Federal Government, in fact, had just four categories of animals: (1) predators, like the wolf and mountain lion, virtually all of which have been hunted or trapped to near extinction; (2) target—i.e., "game"—animals, deer, antelope, bears, etc., which are hunted in regular seasons; (3) "varmints"—coyotes, foxes, raccoons, skunks, etc., which are killed all year 'round, by day and night, by hunting, trapping, poisoning, and every other way; and finally (4) endangered species—animals whose whole population is so low that, under certain conditions, they are protected. Not so much as a word about wild horses and burros.

At least after Annie there were, at long last, two categories of animals who finally did have Federal protection, the wild horse and the wild burro. But the problem was far from solved. Those of us who attempted to carry on Annie's fight for her horses and burros were often glad she had not lived to see the outrageous way both state and Federal authorities went about pretending to carry out what she had started. The Fund for Animals, for example, was during one period actually in court with the Bureau of Land Management for a dozen years—not only over the cruelty of their wild horse roundups but also over the really blatant hypocrisy of their adoption program.

Although the Bureau was, by law, supposed to allow no one person to adopt any more than four horses or burros, over and over again the Fund found out that they had permitted what often seemed like a whole herd to be adopted by just one person—a person who often, after the required one year's holding time, would then turn around and sell their animals for slaughter. On top of this, the Fund several times caught the Bureau red-handed permitting cronies of theirs to operate their holding corrals after their roundups. One such crony, we found out, received $25,000 a day of Bureau of Land Management money for operating such a corral. When I personally visited his corrals and asked to see what animals he had for adoption, he showed me a very small number of reasonably well cared-for animals. After inspecting these, I looked him in the eye and told him I wished to see the cripples. "Oh," he said, "you wouldn't want to see those." I said I did, at which point he looked both surprised and obviously uneasy that I would even know there were such things. Then when he saw I meant what I said, he did everything he could to stop me from going into the cripples' corral. It was one of the saddest and most infuriating sights I have ever seen in animal work, and while looking at them I remember thinking just how Wild Horse Annie felt that day she first saw those wild horses in the truck.

IN THE WAY NONE OF US EVER FORGOT OUR FIRST sight of Peg, the three-legged cat, our people at the Ranch would never forget their first sight of the first wild horses who came to Black Beauty. Early on everyone noticed how remarkably diverse in appearance the mustangs were, not only from other horses, but also even from one another. Many of them had a definite Indian horse look to them—around the eyes particularly. Indeed, one of the most notable of that first group was a special type of Spanish mustang who had bonnet and shield markings over their eyes and were believed, by the Cheyenne Indians, to be magical. One thing, however, was certain, and this was that to any of us who had taken part in anything having to do with mustang rescues, there was magic in all our new residents, if for no other reason than that they would no longer ever be subject to the cruel roundups and adoptions which so often ended, sooner or later, in trips to the slaughterhouse.

Chris Byrne, the longtime manager of Black Beauty, has an extraordinary knowledge of, understanding of, and ability with all kinds of horses. Born in Wimbledon, England, he never succumbed to the lure of being a tennis ball boy but instead, at the age of six, followed a Gypsy woman who carried and sold firewood in a cart pulled by a Welsh pony. Chris became so enamored of the pony that he prevailed upon the Gypsy woman to agree that if he would deliver the firewood, she would let him look after the pony. Chris went on from there to all kinds of animal work and, in the years before he came to Black Beauty, looked after not only farms full of horses but also, specifically, the horses of the Du Pont family. Like Wild Horse Annie, however, Chris had never seen a wild horse before he came to Black Beauty. But, from the moment he saw one he became just as enamored as he had been with that Gypsy woman's pony. Chris particularly loves talking to horses, and all kinds of horses seem equally to enjoy his talking to them. He tries his best not to play any favorites at the Ranch, but he has a hard time disguising that among the mustangs, his favorite is Chieftain.

Chieftain is a white stallion. Only, Chris will tell you, to horsemen there is no such thing as a white horse. There are just different colors of gray, until there is hardly any gray at all and they are what, to a horse neophyte, looks white. Chieftain is actually a dappled gray but, whatever his color, one thing is certain—he is the big chief not only of all the wild horses on the Ranch but of most of the domestic ones too. He loves best to run his herd at least once a day at full gallop, and few sights at the Ranch can compare to watching Chieftain run his herd off in the far pastures just at sunset. Sometimes after such a run Chieftain will allow Chris to approach him and have one of their man-to-man talks. I asked Chris once what they talked about, and he said what they talked about most of the time was how lucky Chieftain was to be away from the roundups and the slaughterhouses. I asked Chris if Chieftain agreed. "He does indeed," Chris said, "in so many words, you might say."

Second only as a talker to the horses was Mary Da Ros, a woman who worked at the Ranch as assistant manager for many years. Besides talking to them, however, she also wrote memorably about them. My favorite among these writings was one she wrote about one of Black Beauty's first mustang arrivals—a filly she named Missy:

OF MISSY

Wild caught by the Bureau of Land Management, Missy came to Black Beauty Ranch from Nevada. Her dark eyes, reckless mane and defiant face made her look strong and earthy, as if she had learned self-reliance while being brushed and groomed by desert wind.

Now, having just arrived at the Ranch and still in the trailer, she was breaking out in a sweat from travel and panic. It did no good to tell her we meant no harm, that this was her home now and she would be able to live out her life in spacious, peaceful pastures. She would have to learn that in time. And with all her pawing and pacing and snorting and stomping, we knew that this would be a challenge, even before we saw the gnarled limb.

When Missy backed out onto the grass we saw, on her right front leg, a knot the size of a melon where her knee should have been. As wild animals instinctively do, she pretended not to hurt, and possibly hoped we wouldn't notice as she limped out. But the deformation made the leg almost unstable as it twisted in a 45 degree angle under her. Though Missy could slowly bring up the leg from behind as she walked, it was at best awkward for her and it was certainly painful for us to watch. Now we knew why no one had wanted to adopt her, why she was here.

Because Missy was so wild and fearful, we had to have our vet anesthetize her so he could examine the knee. After looking her over, he said he could do nothing for it. There were no fluids to be drained, no foreign objects to be removed. The ball at her knee, almost 30 inches in diameter, had, over the years, hardened to a mass of solid calcium.

We considered putting her down. How could we justify her

pain and suffering if she could not live a decent and comfortable life? It is not our policy to prolong the life of an animal with endless drugs and surgeries if there is no hope of a good recovery. We tried looking at other angles, but came up blank. Nothing was going to change the leg.

Except, possibly, if we could get her to lose some weight. We thought if she could shed 100 lbs, it would take pressure off the knee and make her living more comfortable. She was a bit overweight anyway, we commented, even for having survived in a barren desert.

It was then we realized the mare was in foal. For all the attention we put on her leg we had, up to now, overlooked that she was obviously well along in a pregnancy. Our vet tried to continue his examination. But Missy was getting up, albeit wobbly, and obviously wanted nothing to do with us. "Examination over!" she declared. Determining the remaining length of her term was now out of the question. She was moved to a barn where we could watch her and where she had access to a corral so she could walk around.

Several days later, in the quiet cool of the morning, Missy gave birth. When we first saw the filly, she was already up and dry, standing proud with long eager legs. She was solid black with a white star and her eyes, dark like her mother's, were as wild as wind.

Thankfully, this young horse will never have to know that threatened wilderness in which her mother had lived. She will never have to know the endless search for water in an empty pond or a blade of grass through an overgrazed field. And she will never know what it is to be hunted by those who think they have a better use for the land, or a better use for the horse.

Instead, with Missy, she will graze in our lush green pastures. With the other horses she will swim in the fresh deep ponds and run to her heart's content. And she will know what it is to sleep peacefully under the stars, wrapped in the blanket of a warm summer night.

Now we watched Missy lovingly care for her foal, and we could see that she was already moving more easily. With almost 100 lbs of weight already off, she was well on her way to an improved life. And at that moment she certainly wasn't concerned about her leg.

It would be good to know that, as time went on, things got better for the wild horses. It would, however, not be the truth. If the various Park Services with which the

Fund for Animals has had the doubtful privilege of dealing over the years were not enough, the Bureau of Land Management with which the Fund had to deal over the wild horses was even worse. Indeed, comparing the two groups over the years, one with the other, I would have no hesitation in saying that while the Park Services have been, generally speaking, cowardly and cruel, the Bureau of Land Management has been crooked and cruel.

The Bureau of Land Management lost a suit brought many years ago by the Fund for Animals and the Animal Protection Institute—one that required the BLM not to adopt out more than four wild horses to a customer. Here the BLM not only failed in its appeal but also had to listen, in its failed appeal, to the Ninth Circuit Court Judge term the whole BLM adoption program a "farce." And as if this was not enough, as recently as January 1997, Martha Mendoza, of the Associated Press, wrote that more than two hundred current BLM employees had themselves adopted over six hundred wild horses and "could not account for the whereabouts of their animals." Others barefacedly acknowledged that some of their horses had been sent to slaughter. Ms.

Mendoza also wrote that in the twenty-five years since the "Wild Horse Annie" law was passed, the BLM had spent $250 million rounding up 165,635 animals and went on to quote Tom Pogacnik, director of the Bureau's Wild Horse and Burro Program, who conceded that 90 percent of the animals went to slaughter each year, while another agent made what Ms. Mendoza called a "tacit admission" of backdating documents used in the Wild Horse Program.

In other words, the BLM, not satisfied with taking in multimillions of dollars in the continuation of their cruel roundup, and at the same time allowing thousands upon thousands of animals to be slaughtered, went on from these atrocities to allowing its own employees to profit from the slaughter. Even the Department of Justice got into the act, having prepared a memorandum detailing numerous deliberate acts in violation of the letter and spirit of the original injunction brought against the BLM by the Fund for Animals. As an example, the Department of Justice memorandum cited a situation in which the BLM gave title to a horse despite the fact that, before the title was transferred, the adopter told the

BLM inspector that "the horse was mean," and that she "would be very scared to have it around kids or people with dark skin."

In sad summary, all that has happened in the management of a Federal program intended to protect wild horses has been for the BLM to reassign management of the program from Reno to the BLM's national headquarters in Washington, D.C. Here the new "interim director" of the BLM avowed, in a written statement, that "the animals which are under our protection are cared for properly and are treated humanely." If indeed something comes from this avowal, it will be the first time in the long, crooked, cruel history of one of the most despicable agencies in the entire history of the Federal government. In any case, to help this devoutly to be wished event come to pass, the Fund has, as this book went to press, once more taken the BLM to court.

HARD ON THE HEELS OF OUR WORK FOR THE BELEAGUERED wild horses came our rescue of two very different equine customers. These were the last of a long line of so-called diving horses—horses that had long been part of one of the stupidest and most unfeeling animal acts ever foisted on a public which had to be equally stupid and unfeeling to watch it, let alone enjoy it. This act took place at the famous Atlantic City Steel Pier, along with a wide variety of other equally stupid acts such as trained cats boxing each other and kangaroos boxing people. This latter event, incidentally, had at least one redeeming feature—almost invariably the kangaroo promptly knocked down the human.

The diving-horse act had no such redeeming features. It began when a horse was first prodded up a long runway which led to a platform, sometimes forty-five feet in height and at other times sixty-five feet, above a tank below—one that was just a little over ten feet deep. Before the horse made the "dive," a scantily clad young woman, already having climbed a ladder to the top, reached out as the horse went by her, grabbed a bit of harness around his neck, mounted him, and then urged him to jump free of the short ramp leading down to that tank far below.

The idea for this idiocy was originated by a curious Buffalo Bill–wannabe character—a man who had, in fact, a show business acquaintanceship with the late, unlamented buffalo assassin. For years Dr. W. F. Carver, as he was known—although few people ever learned the authority for the "Dr."—had made something of a carnival name for himself by his prowess as a sharpshooter, his most memorable stunt being his ability, by his own well-trumped-up figuring, to register in skeet shooting 10,000 hits a day for six days running.

One day, on a trip to Australia, Dr. Carver persuaded two of the Buffalo Bill troupe playwrights to write a play in which the central character would be played, not by Bill, but by him. In this there was a scene in which Dr. Carver had to ride a horse over a bridge over a river. Midway in this effort he had a lever which was pulled by a stagehand which collapsed the bridge, which in turn collapsed into something looking, at least from the audience, like a river. When these collapses came, Dr. Carver jumped up and caught hold of an upright, thus masterfully upstaging—something at which he was remarkably adept—not only his rival Buffalo Bill but also his horse, who had to plunge down, and disappear, into the aforesaid river.

The only problem the playwrights faced with their absurd drama was that once a horse had to fall off the bridge and into the river, that same horse was so loath to repeat the scene the following night that each night a new horse had to be procured for the job. In short order the company ran out of horses, and the whole show would have been brought to a well-deserved final curtain had not Dr. Carver resolved to find the answer to the problem by enlisting one of his own horses, one named Silver King, who apparently did not mind being goaded into falling into the river night after night. In no time at all Dr. Carver, having gotten rid of such encumbrances as Buffalo Bill and the playwrights, was well on his way to perfecting a brand-new act—a diving horse, courtesy of Silver King and the scantily clad young woman.

Actually, as the diving-horse act grew in popularity, Dr. Carver's son, Al, soon produced a second company, which went national. From time to time both Carver and his son cut expenses by letting their horses dive without even bothering with a girl on top. In the early days, too, he occasionally ran afoul of some spoilsports who felt the act was cruel. So weak, though, were the animal groups in those days that they failed to achieve their objective to make headway against the Carvers. On one occasion, however, an animal group in California managed to take Al Carver's company to court, whereupon Al proceeded to put one of his horses in a truck which he himself drove all over town and on which was a large sign over the horse which read, "I'm Being Taken to Jail for Jumping in a Tank of Water." Afterwards, during the court hearing on the case, Al attempted to take his horse into court to show the judge what good shape he was in and was only persuaded not to do so because his crew warned him of the danger of taking a horse—which was not shod, for the diving act—up the marble steps to the courthouse. Even with the horse kept outside on the lawn, however, the Carvers won their case easily. The judge recessed the trial, came out of the courthouse himself, inspected the horse, and threw the case out of court.

Despite the "victory" for the Carvers, there was of course abundant cruelty in, if not the act itself, the whole training program for the diving horses—the same way as there almost always was in the training of so many other animals for such idiotic acts as dancing bears. There was also danger to the girl in the act. The horse had no saddle, no bridle, and no stirrups, and not even a bit in his or her teeth. Instead, the girl had to hold on to guide the horse by two leather bands—one around the horse's neck and the other around his body. Both the kick-off board at the bottom of the horse's take-off ramp and the sides of the platform from which the girl jumped on the horse were padded—the kick-off board so that the horse had something to kick from and the sides so that the girl did not get friction burns.

The diving girls learned the act on a twelve-foot platform rigged beside the real high one. They were taught that when the horse first put his hooves over the edge of the down ramp, they would have the feeling that the horse was going to somersault and that they would go off over his head. However, if, just before the horse got to that kick-off point, the girls leaned back and kept a firm hold on the harness, they would not be injured. Nonetheless, on one occasion a girl in training got so frightened she let go of the harness, and then of course shot right off the back of the horse. Going off like that, she did not even reach the tank of water—instead she landed in front of the tank on the bare floor, and was only saved from breaking her neck by the fact that she landed absolutely flat.

Actually, the landing in the tank was no walk in the

park. Sonora Carver, the Carvers' most famous girl rider and the wife of Al, found this out one day when the horse she was riding made an almost straight-down nose-dive. She recalled the event in her book, *A Girl and Five Brave Horses*:

Normally I would have ducked just before hitting the water and would have entered the tank foremost, but this time his body was in such an extremely perpendicular position that I was afraid of throwing him over on his back. Several times before on a nosedive I had struck the water in an upside-down position, a method of landing that can be both painful and dangerous, for even if a rider escapes without serious injury she gets a severe shaking up. I had to make a split-second decision, and in a desperate effort to avoid turning him over I stiffened my arms and held my weight back, hoping to maintain our balance. I was successful, but that position caused me to strike the water flat on my face instead of diving in on the top of my head. In the excitement of the moment I failed to close my eyes quickly enough, and as we hit the water I felt a dull stinging sensation.

After the experience, Mrs. Carver first put off going to an eye doctor for some time and then only did so after another experience in which she hit the water with her eyes opened. It was then, however, too late, and she eventually went totally blind. Remarkably enough, even totally blind she kept on diving horses for eleven years, wearing a helmet with a special lens to protect her eyes. Almost equally remarkably, for the first five of those eleven years she kept the fact that she was blind from both the press and the public. During this period, however, she had many close calls with other accidents, either by missing mounting properly as the horse entered the platform preparatory to the jump, or by losing her grip on the harness and getting thrown off when the horse hit the water. Despite this she managed to get by without serious incident until she retired.

On the other hand, not so lucky were many of the other diving horse girls, some of whom were badly injured. Sonora's sister, Arnette, who also had a long career of diving, told me that it was by no means easy to grab the horse's harness from the platform. When she was learning she sometimes missed the harness entirely, after which, she remembered, the instructor made her, as well as the horse, do it again. "I noticed," she said, "it always made the horse very mad to have to do it again." Even after she learned to dive regularly, she also learned that some of the horses would roll over when they hit the water—at which time, she said, "you had to help them get right-side-up for swimming."

The horses, of course, also suffered accidents. Indeed the most tragic of these happened to the most beautiful and skillful of all diving horses—one named The Duchess of Lightning. Always trying to stretch the act to the limit, the authorities in Atlantic City decided that rather than have the horses dive into a tank, it would be more amusing to have them dive into the real ocean. Accordingly, one of the Carver contracts called for The Duchess of Lightning to do just this. The Duchess had done the ocean dive before at Atlantic City, but the new contract called for her to do it at a seaside resort in California. Here, unfortunately, the conditions were different. On her first attempt The Duchess apparently made a beautiful dive, but afterwards, turning to swim ashore, she became confused by the shore breakers and, in attempting to find an easier way, got herself turned around and headed for the open ocean. Seeing this, lifeguards jumped in a boat and started after her. The faster they went, however, the faster Duchess swam away from them until finally, with the lifeguards watching from their boat, in the words of Al Carver, "She gave up—put her head down and drowned."

That should have been enough to end the idiocy of the diving horses for good, but it was not. Year after year the act was renewed at both Atlantic City and other resorts. And although the animal groups grew stronger and stronger in opposition, they were not enough to put an end to the act until the 1980s. Even after this, and long after the diving act was stopped, the Disney studios not only did not criticize the act but instead seemed to support it with a pro-diving-act movie about the blind Sonora Carver—one which bore the curious title *Wild Hearts Can't Be Broken*. Fortunately for the by-now vast majority of anti–diving horse people, the movie was a well-deserved failure—in fact, it was one of the largest flops in Disney history.

As Fund for Animals workers as well as others will attest, not only was it difficult to stop the diving-horse act; even after it was stopped, there were still other similar acts regularly doing business. The most persistent of these was Johnny Rivers and His Diving Mules. Although they dove from a relatively low height, the mule act was particularly belittling to animals because it also involved monkeys tied on top of the mules and even monkeys tied on top of dogs who were tied on top of the mules. According to two of the longtime diving-horse trainers, Ruth and Bill Ditty, there was even for some time an act involving diving zebras. "They didn't go over too good," the Dittys reported to me. "They were awful high-strung and hard to handle." I could well understand why.

PEOPLE WHO DO NOT KNOW HORSES THINK OF THEM as big and strong and healthy. People who know them, however, quickly learn that they are extremely delicate creatures and are highly prone to disease and all kinds of accidents. At the same time, as the patient animal slaves they are to mankind they suffer all kinds of cruelties at mankind's hands. Start with the outrages in horse racing—from the racing of two-year-olds and the drugging at the racetrack, and go on to distance racing, steeplechasing, and the horrible "Grand National," not to mention such "genteel" sports as polo. In all these the horse has a grim time indeed. Outside of sports the menu of misery for horses is a long one and goes all the way from carriage horses plodding the city streets in blazing heat and traffic, to Tennessee walking horses whose training cruelties pass belief, all the way to *Sports Illustrated*'s recent report on the incredibly brutal bludgeoning of show horses for the insurance money. Placed against such examples it is not surprising that such a seemingly minor matter as diving horses should take such a long time to overcome. It received such wide publicity, however, that it became essential for something to be done about it. The Fund for Animals has always felt that not doing anything about highly publicized cruelties is wrong for many reasons, not the least of which being that, if nothing was done about them, then people would either think that the act wasn't that cruel, or perhaps even that the horses enjoyed it.

In any case, there was a strange postscript to our efforts. Out of the blue one day came a letter to the Fund from a woman who told us that Resorts International, the outfit that had bought the Steel Pier, had decided to close down the diving-horse act for good. They were going, she told us, to sell off the last three diving horses at auction. As she said this, we knew that meant they would be headed for the slaughterhouse. Again, it might seem curious, since thousands of horses go to slaughter every day, to try to stop three of them from that fate just because they were diving horses. But once more, the fact that they were going to slaughter would be a highly publicized cruelty and might ultimately affect not only the fate of those horses but perhaps help some others. In any case, almost at once we made a decision to try to get those last three horses and bring them to Black Beauty Ranch.

To achieve this objective we dispatched Cynthia Branigan, a longtime Fund for Animals field agent from New Hope, Pennsylvania, first to find where the horses were, second to stop them from going to slaughter, and third to try to get them for Black Beauty Ranch. Promptly Ms. Branigan reported first bad news, then medium news, and, third, good news. The bad news was that one of the horses, Powderface, had already been auctioned and gone to slaughter. The medium news was that, because of this, the public was already aware of the selling of Powderface to slaughter and since the Fund for Animals had now entered the picture, the second horse, a twenty-six-year-old big, dark bay named Gamal, would now go for a large price. The third, last, and good piece of news was that the third horse, a seven-year-old small chestnut filly named Shiloh, had already been bought at auction and was owned by someone, but that person was willing to sell her.

Only on very special occasions and for very good reasons does the Fund for Animals pay for animals at rescues. And on this occasion we decided that before we went after Gamal and Shiloh we would ask Resorts International if they would like to help in the rescue. At first they were happy to do so and even admitted that the reason for this was that they were tired of receiving bad publicity just for having owned the diving-horse act, even if they had now stopped it, and felt that helping with this rescue would at least counter that, and get them some good publicity. Unfortunately, just before they did so, they suddenly changed their minds and said they would not help us. When we asked their reason for this, they said they now felt that any publicity connected with the diving horses—even an attempt to rescue one—would be detrimental.

In one way this pleased us—it proved at last that by far the majority of people were that much against the diving-horse act. But in another way it did not please us, because we couldn't help wondering whether the people who didn't like the idea of Resorts International being connected to their rescue only hesitated because the diving horses were stars and they didn't like the idea of their being reduced to hamburger. Whatever the reason, the upshot left Cynthia alone to face the job of somehow getting ahold of the last two horses, at least to avert the possibility of the slaughterhouse. In the end she did it, managing to procure Gamal for the Fund for Animals for $2,600. The woman who had bought Shiloh, however, was now not willing to resell her. In view of all the publicity, she realized a good thing when she had it, and before it was all over Cynthia had to pay more than twice the price the woman had originally paid for Shiloh.

On top of all this, our plans to take both Gamal and Shiloh to Black Beauty went awry. The reason for this had nothing to do with Shiloh, who still today is an extremely happy Black Beauty horse. It had to do with Gamal, or rather Cynthia and Gamal. Cynthia, like so many women I have known, was unreliable where the first horse of her very own was concerned. In the two days she had Gamal before he was to be taken off to Black Beauty, she fell head over heels in love with him. Above all, she wanted him to stay with her and not go all the way to Texas. To build her case she foisted argument after argument upon us, claiming she would pay for every cent of his stable care and would look after him every day. She even went to such despicable lengths as persuading various horse authorities to tell us that Gamal was far too tired and sick to make the trip to Texas. As if that were not enough, even after it was obvious that Gamal was neither tired nor sick she pressured the same authorities into telling us that he was now too old to make the trip.

Actually, Gamal lived for ten years under Cynthia's faithful care. "I loved him," she told us, "not because he was a beautiful, sleek, quivering thoroughbred. He wasn't anything like that. He never quivered at anything—he was as tough as nails. In all those ten years I only rode him once, and then just because I wanted the experience. Almost every day I would go over to his pasture and walk around with him and talk to him and hug him and kiss him. I guess you could say it was just one more of that 'little girl and big horse' thing. Anyway, I know he liked me because I was a girl, and I liked him because he was a boy."

One day Cynthia and I took Gamal for a walk down to the Delaware River. Gamal went right into the water, up to his chest, and then sank down and plopped back and forth. As I watched this, a horribly disagreeable thought entered my head. You don't suppose, do you, I said darkly to Cynthia, he misses diving? For some time Cynthia, too, watched what he was doing, then finally answered my question. "No," she said, "I don't. And, even if I did, I promise you my lips will be sealed."

OUR DOMESTIC HORSES WERE PERHAPS NOT AS EXCITing to other people as our wild horses or our diving horses, but they were equally exciting, and often more so, to us. Two very recent arrivals who stand out are two huge, beautiful Belgians—sister draft horses with especially beautiful fur over their huge hooves. Feathers, these are called. Told by an informant that the horses had no place to go, having first been workhorses in Amish country and later in a commune which was now about to be abandoned, we immediately offered to take them at Black Beauty.

There were, however, as occurs so often in our rescues, three difficulties. The first was that the horses needed immediate help and medical attention, the second was that they were hundreds of miles away, and the third was that the weather, typical of Texas, was a combination of driving rain and incredibly long-lasting thunderstorms. None of these matters bothered Chris Byrne in the slightest. He was off in our truck and trailer almost immediately and, when he got to where the horses were and realized that the weather had, if possible, worsened, he reacted in typical Chris Byrne rescue fashion and decided to stay overnight right with the horses in the trailer so that, as he put it, "I could get to know them better."

"I guess what I like best about horses," Chris once told me, "is just watching the way they interact with each other." And certainly that night, with those two huge horses in the trailer with him, he had plenty of opportunities to do his watching. By the time he was back at the Ranch it was clear he had a special feeling for Hobby and Polly, as the giant sisters were called. There was still much work to be done, particularly on their hooves with those beautiful feathers. All in all, because he knew they needed work but also because I had a sneaking suspicion he just liked them that much, we decided to keep Hobby and Polly right down near where we keep all our old and infirm animals, what we call our "house animals"—as near to the main ranch house door as possible, so we can keep a close weather eye on them.

Both these horses needed Chris's very firm weather eye, because in their former situation they had been both overworked and undervetted. When I spoke earlier about horses being often thought of by nonhorse people to be strong—as in the expression "strong as a horse"—Hobby and Polly clearly proved that not only was that expression generally wrong, it was even wrong when it came to two giants like Polly and Hobby. Both had in fact, before they came to us, foundered. Once a horse has foundered it can often founder again, no matter how big and strong he or she appears, and can do so even from such a simple thing as having eaten a too-rich diet. From this, indeed, it is even possible for a horse to die.

Chris was particularly concerned with the horses' hooves, which had apparently been severely neglected for years and would need much and regular attention. During this period the huge horses spent much of their time with the very smallest horse at the Ranch, a crippled miniature horse named Pee Wee. At just three feet or, in horse talk, nine hands high, Pee Wee could literally walk back and forth under Hobby and Polly. And the sight of the three of them, soon all close friends, walking around and, in Pee Wee's case, under, one another, was a very special one at the Ranch. It was particularly so because Pee Wee's crippled shoulder bothers him so badly that he is normally inordinately shy about any larger animal, be it four legged or two legged, and only made this extraordinarily large exception to his shyness because of his complete confidence in his firm friendship with Hobby and Polly. I have no doubt, however, that Pee Wee, one of the smartest horses on the Ranch, also immediately realized that his two newfound friends would also make, in time of trouble, perfect bouncers.

Looking back on all our domestic horses that have gone through the gates of Black Beauty, one that still stands out as special in the minds of all of us is a beautiful mare named Jamie. Mary Da Ros, who wrote the moving saga of Missy, also wrote the saga of Jamie:

OF JAMIE

What I remember most about Jamie were her eyes. They were graceful and brown like a deer's, only bigger. They were the kind of eyes that, when you looked into them, gave you an immediate connection to Jamie. Although, of course, they did not tell you everything—not, anyway, everything she had gone through—they did tell you what kind of a horse she was, and how, in spite of everything, she still loved people.

Jamie was a horse who did, without hesitation, all that was asked of her. She made the perfect brood mare and, for years, she delivered beautiful foals for the various people who owned her. She had the run of the pastures and she used them to raise her young—where they, like she, could run to their hearts' content.

That is, she had their run until she was crippled by a carelessly discarded piece of metal. She tripped on this and was badly cut just above her coronet band. The coronet is the growing part of the hoof and extremely sensitive—and, unless the horse has a great deal of luck, and the immediate attention of a very competent veterinarian, any horse sustaining an injury to the coronet will have a permanently injured hoof. Jamie it seems was, at the very least, out of luck.

Although she was treated and put on medication, her hoof

never healed. She went lame. It would seem that this would have been a good time to retire Jamie from the horse-breeding business. But her owners thought otherwise. They kept her on heavy doses of pain medication so that she could continue to produce foals which they, in turn, could sell. That, apparently, was the important thing.

Eventually, even with all the pain medication, Jamie could no longer bear her own weight and was unable to stand for long periods of time. Her foals learned to nurse on their knees, taking milk from their mother while she was lying down. Finally she could not even do that, and that was when she came to Black Beauty Ranch—when her owners had finally sought to retire their horse and had given in to their conscience. Evidently they did have one, small as it was.

Over the phone they told us that Jamie had a leg injury and with continuing medication she would eventually heal. They wanted to send her to the Ranch so that she could, eventually, run free with our wild mustangs. "Nothing but the best," they said, for their horse. Evidently their conscience was now working overtime. Ever suspicious of people who want to retire their horse once it is no longer "useful," and unable to hear the details from Jamie's side of the story, we could only guess at how awful it really was. We said of course we would take her immediately.

When Jamie arrived at the Ranch we found her to be far worse off than we had been told. She had little to no use of her front legs and moving at all was incredibly painful for her. In fact, in order for her even to turn around she had to place all of the weight from her front legs onto her back legs and then, picking both front legs up off the ground, used her back legs to do all the work and slide herself around. Because horses normally keep most of their weight towards the front, she had developed severe and frequent muscle spasms from all of the weight shifting she had to do. To watch her was nothing short of agonizing.

It is customary to have all of the new horses arriving at the Ranch examined by our veterinarian and certainly there would be no exception here. When Jamie first arrived we knew that she had been living with this frightful injury for years—it was a very old injury, and a very serious one at that. This was all confirmed by Jamie's sad, tired eyes and it was reconfirmed again by our vet. But that was not all. We also found out that Jamie had developed an ulcer from all the pain medication she had had to take. And now we had to give her more drugs for the ulcer, and yet she still could not go off of the pain medication.

For months we tried to help Jamie with her condition, and with every treatment we gave, we grew more attached to and fond of her. We liked to think that because she was now at Black Beauty, where our objective is to do what we can for the horse and not what the horse can do for us—she started to relax, and she started to trust. She seemed to be trying so hard to get better and helped us to administer her treatment in every way she could, never flinching and always brave. Her big brown eyes lit up whenever we came toward her and I felt them follow us when we left. And when she saw the other horses running free in the outside pastures I was sure she wanted badly to join them. She whinnied at them often.

We tried everything short of surgery to help Jamie with her condition. Finally, after watching her remain the same, despite everything, and still being caught in a serious tangle of drugs, we decided to have surgery done to block the nerves in her front hoof. Then she could use it, but she would not feel the pain. And, if everything went well, she could come off all medication and live a somewhat normal life. Optimistically, we tried to picture the day when she would be able to run in the pastures again.

The day before Jamie's surgery, we loaded her carefully into a horse trailer to be driven to the vet's office. After parking, I opened one of the trailer windows and she put her head out to get some fresh air, and get a grip on where she was. As I walked toward the vet's office, she whinnied. I looked back and there were those big brown eyes, soft and trusting as always, looking into mine.

Although that was not the last time I saw Jamie alive, it is that moment I will carry with me and remember always. With her beautiful head sticking out of the trailer window and watching every move I made. That is how I will remember Jamie.

The surgery did not go well and there were complications due to the serious ulcer that had developed over the years. Upon hearing the news we rushed over to see her. Although she was still alive, I could see that Jamie was already gone. Lying there was an animal in so much pain who could not possibly recognize—did not want to recognize—what was going on around her and that we were there for her, even though we could do nothing to make it stop.

We rushed off for the veterinarian and had her put to sleep immediately. I held her head and then it was over. We were so hurried to put her out of her misery that we did not even pause to say good-bye. But after it was over we sat quietly and softly said our good-byes to the Jamie we remembered and the Jamie we would like to remember—Jamie in the pasture, Jamie in the stable, Jamie running in the field with her foals. And then there was the Jamie sticking her head out of the trailer window—and then, finally, there were just those eyes. There was Jamie.

We have been lucky not to have had many, if any, horse stories as sad as Jamie's—although the majority of our hundreds of horses were either abused or ill-used to begin with. One horse who particularly comes to mind in this regard was a small colt all alone in a field to whom we were directed by someone who came upon him by accident. It was a starvation case of which there is an inexcusably large number in Texas, but this one was particularly inexcusable. When the rescuer first saw him he was trying to make a pathetic little whinny. He was lying in the dirt, cold and wet and barely alive, beside a tree. He was also, we could see from the dirt around him, trying to make his way from a lying position toward the tree.

The day we found him was, ironically, Thanksgiving. How long he had been where he was, how long without water or food, how long without his mother—where she was or even if she was—we would never know. All we did know was that someone had left this tiny little colt to die a long, lonely, painful death and we also knew that the most important thing we could do for him was to give him water and something for his Thanksgiving dinner, besides the bark of the tree he was trying to reach. We also decided, once we had him safely in a stall in the main Black Beauty barn, that we would name him, in honor of the holiday, Pilgrim.

The veterinarian gave Pilgrim less than a ten percent chance of survival. But Pilgrim was not a colt who gave up. He just, as the vet later said, refused to consider that option. Pilgrim never grew up to be a large horse, or even a medium-sized horse but, as he had already tried to prove, he did grow up to be a brave horse. And, after months of medical care on the vet's part and that special bravery on his part, he started to come around faster than anyone thought he ever had a chance of doing. And, as he did so, he began to take an interest in the other animals at the Ranch. He eventually even tried to make a four-legged friend—something, it seemed to us, he had never had before his rescuer found him. In any case, this four-legged friend turned out to be, of all the horses on the Ranch, Shiloh, the last of the diving horses. It was certainly eminently fitting, for Shiloh, in her way, was a brave horse, too. And, even more eminently fitting, a year and a half after their meeting Shiloh had a foal, and Pilgrim was a pilgrim father.

Of all the Ranch stories of man's inhumanity to horse, the story of a horse named Cody is not only, like Pilgrim's, inexcusable, it is close to infuriating even to think about. And although it started out, again like Pilgrim's story, as if it could not possibly end happily, in the end it did, or at least as happily as possible. Cody, a beautiful white or, in horse talk, gray, was owned by a doctor in Atlanta—a doctor who, one day, shot him. He shot him in the knee, the reason being that, although Cody would not come to him when he called, he would come to the young man who looked after him. The doctor, in a fit of fury over this, and jealous of the young man, was not only not content with shooting Cody in the knee, he had such a fit of fury that he fired the young man, then rigged a block and tackle with weights and left Cody so shackled with a wound so unattended that there was no hope for even partial recovery. In other words, he wanted to see that Cody would not only be crippled for life but would also be tortured for life. All for not coming when he was called.

It was not all, however, for a group of Fund for Animals volunteers in Atlanta. Learning about this story from the young man, they took the doctor to court about Cody and, although they were able to get him fined for cruelty, try as they did they were unable to get the judge to accede to their demand that Cody be taken away from the doctor and that they be given custody of him. Or, if not that, that at least he should be taken away from the doctor. As happens in all too many cruelty cases, the judge refused to do this. The volunteers did, however, succeed in making the doctor so infamous in Atlanta that in time he gave in—or, rather, gave up—and sold Cody at auction, a fate that might well have led, as the doctor might well have hoped, to the slaughterhouse. But this time the volunteers were ready for him. They not only bought Cody at auction but, besides this, also saw that he had a proper operation on his knee. In this operation—which, incidentally, took more than four hours—surgeons first removed the bullet from Cody's leg, then fused the splintered bones back together, pinned the knee with stainless steel screws, and finally implanted a metal plate to support the fractured bone. After all this, the volunteers sent Cody to our Ranch.

To this day a fragment of the bullet remains in Cody's knee, a fragment that was unreachable due to its closeness to the nerves and an artery. As I write this Cody is now twenty-seven years old and has been living at the Ranch for more than fifteen years. He is doing well and although he doesn't gallop anymore, he has been known to at least try to trot a little bit, especially if it is to the trough at feeding time. Cody's toughness earned him the respect of all the other horses, particularly in his early days, when he was lamest, when a horse named Stanley took it upon himself to stand beside him at feeding time and see that there was no nonsense about who got what, especially when the who was Cody. Our farrier made a special shoe for Cody which allows him to walk a little easier, particularly as he grows older and his bones become weaker. Just the same, to be on the safe side we have always kept Cody as near to the main Ranch house as possible, just because we want him close. And although he is always nearby, he is just a little out of reach because he remains, to this day, to most people, people-shy. He is even a little shy with those of us he has learned over the years to trust. We believe that, given his experience with people, he will always be so and we highly respect him for it.

There remains, finally, the story of Stanley, a grand sorrel gelding. Stanley has passed on now, but no animal at the Ranch was ever more loved or ever more missed. Most of his years were spent as a police horse on the streets of New York City—in the heat in the summer, in the cold in the winter, in traffic in rush hours, in unruly crowds, in crimes and on concrete. Wherever he was, and whatever he did, Stanley, always the gentleman, did his job with pride. And when his duty was finally over, and he was sent by friends to retire at Black Beauty, he saw he had a different job, but one that should be done with the same pride with which he had done his former job. This one did not involve policing a whole city, but it did involve policing a very large ranch and six hundred animals, and among these, when he was there, he made sure that there was no bullying by larger animals of smaller animals and no stealing of anyone else's food.

People remember different things about Stanley—how big and strong he was, yet how gentle and friendly, even to another animal who, at least at first, was not that way to him. But they always were that way in the end because Stanley had more friends at the Ranch than any other animal ever had before or since. His best friend was Cody. He stood by Cody's side through thick and thin, especially watching out for him against some of the other horses who, knowing Cody could not move very well, took advantage and, at least before Stanley, used to take his food. When Stanley took over no other horse, not even the wildest mustang, ever even tried to do such a thing. And when Cody was, as he was so often, sick and had to be in the barn, Stanley would wait for him outside day after day. Then finally, when Cody would emerge, they whinnied, nickered, and went off grazing together, business as usual.

Next to Cody, Stanley loved best the burros. He searched them out to stand by them, and would go to great lengths for them as he would for Cody and his other friends. He would slowly make his way from the stables, around the elephant barn and corral, and then down the lane to wait for someone to walk or drive to the gate so he could be with his burros out on the hill. Eventually he even learned to walk gingerly across the cattle guard—a difficult feat, as he grew older—so he could get to the burros on his own.

Anyone who conducted a tour of the Ranch would never forget how he or she could always count on Stanley to be so steady when children rushed him. And while it would have been a little on the dangerous side to see so many small children so close to almost any other horse, it was not that way with him. Even the most nervous mother could see immediately that her smallest child would be safe with Stanley. Indeed, when one of those very small ones would want to touch him, Stanley would put his nose way down so that that very small child could touch his soft muzzle. All the children loved Stanley very much—perhaps more than any other animal at the Ranch. But just the same, those of us who worked with him loved him the most.

Stanley is buried now—out on that hill with his burros. We believe that he will always be in their hearts, too. And one day Cody will be out there with him, and they will always be together.

Chapter Four

OF THE SMARTEST ANIMAL OF THEM ALL

WHAT ANIMAL IS THE SMARTEST? THIS question, often asked and rarely satisfactorily answered, is, to begin with, not easy to answer. The probability is that most people would nominate either the dolphin or the chimpanzee—and certainly I would be inclined to second either candidate if for no other reason than I once spent an unforgettable afternoon swimming with dolphins and never forgot either their brightness or their gentleness.

At the same time, I have had close associations with at least four elephants and I have many times been present when they clearly not only do things that I do not believe either a dolphin or a chimpanzee could do but also think things that I did not believe they could think. And, while we are considering the matter of smartness, how can we forget the Seeing Eye dog or, for that matter, either your cat or, from my point of view, obviously more importantly, my cat?

Then too when it comes to assessing the smartness of animals we should not, in fairness, forget the bird family. I well remember staying at the home of Bill Carlson and Nancy Nelson, two friends of mine who are distinguished television personalities in Minneapolis, being impressed by their whole menagerie of household pets, and also noting that a parrot they had was both boss of all the other animals—which included both dogs and cats—and easily the smartest of all of them. Nancy, noticing my focus on the smartness of the parrot, immediately chided me for not knowing enough about the smartness of birds in general. Indeed the very next evening she took me to a party at the home of a friend of hers where one of the guests, clearly at Nancy's behest, brought a bird and perched it on her shoulder. The bird remained the entire evening. By the end of the evening I was convinced that I had become friends with the smartest creature I had ever, at least up to that time, known. The creature was, of all creatures, a common garden-variety crow.

From that day to this I have never forgotten how smart that crow was—and I do not mean just looking around and taking in what was going on at the party, but also seeming clearly to understand what was being said. In any case, ever since that experience with the crow I have had qualms about stating what animal, or for that matter vegetable or mineral, is smarter than any other. I shall still state unequivocally, however, that at Black Beauty Ranch the smartest animal, of the hundreds upon hundreds we have had there, is a chimpanzee.

His name is Nim, and he was, by the time we got him, the most famous chimpanzee in the country, if not the world. The reason for this extraordinary fame is simple—Nim could talk. I do not mean talk in the sense of human talk, but rather in the sense of sign language talk. Nim was the quickest learner of all the "signing chimps," as they have been familiarly called. He was also the closest to making actual sign sentences.

I first learned about Nim when I read an article about him stating that, having been taken away from his mother at an incredibly early age and taken in by a human family, and having studied under the direction of a large corps of researchers and teachers—one article said there were sixty of them—he had now completed his learning and was going to a medical research laboratory run by New York University. There he would, the article said, be the subject of a study of the effects of a hepatitis vaccine which would involve, among other things, his being in total isolation for a period of at least six months.

In other words, having spent years—almost four, in fact—learning to do everything his "family," as well as his researchers, teachers, and students wanted him to do, including learning to be a human and talk human, he was now going back to being a chimpanzee again all right, and to talking chimpanzee, but he was going to be in isolation from other chimpanzees for at least the first six months and would then presumably be a permanent laboratory chimpanzee and be experimented upon for life.

Although I am not one of those people who believes that it is all right to experiment on some animals but it is not all right to experiment on other animals, I found it ironic to the point of cruel absurdity that an animal who had learned to do something no other animal in the world had ever done any better would finally be thanked for it all by being made just one more of the many hundreds of thousands of laboratory primates.

The laboratory where Nim was now imprisoned was a place called LEMSIP, in Sterling Forest, up the Hudson in New York. Nim was born in Oklahoma, at something called the Institute for Primate Studies at the University of Oklahoma. I had no idea what that was and at the time had no idea that the Fund for Animals could, even if it wanted to, ever rescue Nim. But having had some experience in the field of rescuing laboratory animals I was convinced that we would have a better chance bringing publicity to bear on the University of Oklahoma than we would on the LEMSIP laboratory. Almost immediately the Fund began an intensive campaign against the University of Oklahoma, using as much press as we could muster as well as radio and television. At first we got literally nowhere. The University simply ignored our campaign. Gradually, however, as we were joined by other groups, we began to create cracks in their resistance. Finally, one day I got as far as having a personal conversation with a major officer of the University. Up to that time he had had just two arguments against us. The first was that the University of Oklahoma had a contract with LEMSIP and he did not see how he could break it. The second was that even if he could, he wouldn't.

Our answer was to redouble our efforts. I went back to the same officer of the University I had spoken to before. This time, although he stuck to his argument, I felt there was a chink in his armor. He was obviously irritated at the volume of letters and phone calls we had been able to generate—so much so that when he mentioned this I was emboldened to ask him if he thought it was a fair thing he was doing, to send Nim to a laboratory after all he had tried to do for people. He hemmed and hawed but clearly personally did not think it was fair.

I liked to think that my talks were responsible for what happened, but whether they were or not, one day we picked up an article which said that Nim had been returned from the LEMSIP laboratory to the University of Oklahoma, and that he was now back where he had been born, at the Institute for Primate Studies. While we were all celebrating the news, at the same time we realized that this was not yet the end of the Nim story. For one thing, we still did not have the slightest idea what the Institute for Primate Studies was, and for another more important reason, we didn't have the slightest idea what kinds of "studies" they did.

That very day on which we heard the news that Nim had been returned I decided to call the Institute. My call was answered by the Director himself, Dr. William Lemmon. He was clearly aware of the part I had played in getting Nim away from LEMSIP and back to the Institute. But to my great relief he was not distressed by this. All in all, I was again emboldened to ask him a direct question. This was, would it be possible for me to visit the Institute? This time I was even more relieved, especially when he asked me what day I could come. I suggested the very next day.

When I arrived at Dr. Lemmon's office I soon saw that Nim was just one of more than a score of chimpanzees the doctor had at his Institute. While he told me about them, he also told me he had many others who were apparently on an island in the Caribbean. As for the various "studies" to which his chimpanzees were subjected, again I was extremely relieved to learn they were noninvasive—they were behavioral in nature. From what I could glean from Dr. Lemmon's conversation, he seemed to specialize in sexual studies—more than one of which seemed to me, as a Bostonian, extraordinarily personal. An example, which he dwelled upon at some length, was that when he went to bed at night he sometimes put one of his chimpanzees between himself and Mrs. Lemmon in the double bed. The crucial part of this experiment, as I understood it, was how long it took the chimpanzee to push him out of the bed. Dr. Lemmon did not go into any further detail, for which I was both relieved and grateful.

As Dr. Lemmon went on talking about his experiments and his chimpanzees which had taken part in them, I observed that he never mentioned Nim. In fact, the more he talked about all the others except Nim, the more I got the impression that there was something odd about the way he felt about Nim. It dawned on me that although he did not seem to be jealous of Nim personally, he was extremely jealous of Nim's fame. It also became clear to me that, being so full of his own experiments and so proud of them, it was almost natural that he would be annoyed that people would constantly call, write, or even come to see him—as he told me they did—invariably about Nim, not about his experiments. Then and there I resolved to concentrate on every facet of his experiments—boring as they were to me—and not utter so much as a word about Nim.

When lunchtime came, Dr. Lemmon took me to the faculty club at the University and once more launched into his experiments, whereupon once more I launched as much interest as I could. Finally after lunch, when it seemed my visit was over, he suddenly asked me if I wanted to see Nim. I told him, trying not to evince the slightest unseemly enthusiasm, that I would be happy to do so.

Arriving at Nim's cage, I was amazed to see first what a large and powerful animal he was and second how, despite being larger and more powerful than the chimpanzees in the same cage, he hung back, not at all afraid of them but at the same time displaying very little interest in them. It was very clear to me that Nim was one extremely unhappy camper.

Despite this, I still said nothing. Dr. Lemmon made up for my silence by unleashing a full river of conversation about Nim. "There's been so much *fuss* about him," Dr. Lemmon said. "Day and night, telephone calls, people wanting to come and see him, over and over." He paused. "I really don't know what to do about it. Everybody seems to want him, too. You wouldn't believe the offers I've had. Circuses, animal parks, everybody. All the zoos, too. And you'd be surprised what some of those zoos have offered me for him. But I don't like zoos."

As Dr. Lemmon went on talking I kept looking at Nim. I had come to see Dr. Lemmon not to take Nim away from him but just to find out what he was doing with him and hoping he would not again allow him to go to another laboratory. But now, seeing how unhappy he was, I was galvanized into a desire to have him. And I was galvanized not because Nim was so famous, or so intelligent, or so valuable, or so anything else, but simply because he was so unhappy. "The trouble is," Dr. Lemmon said, "he was born a chimpanzee, but now he isn't a chimpanzee anymore. He's half human." Dr. Lemmon paused. "That's what he really is," he said, "half and half."

Patiently I waited until Dr. Lemmon finished. Then it was my turn for a river of conversation. I told Dr. Lemmon why we had built Black Beauty Ranch and why, in my opinion, it would be a perfect home for Nim. I told him that I would try to persuade one of the Fund for Animals' donors to build Nim a special home. I told him that I'd always thought of Black Beauty Ranch as a kind of Noah's Ark, and that we always liked to have at least two of any new kind of animal. I even asked him if he would go so far as to pick out a special female for Nim—one of over breeding age, I added sternly—to be his companion. As we said good-bye, Dr. Lemmon did not say anything about agreeing to let the Fund have Nim and a friend, but at the same time at least I knew he was considering it.

Somehow, by the time I left Dr. Lemmon I was so sure that he would go for my idea of the Fund taking Nim that I went immediately back to the Ranch and started calling donors about the idea. I did not have to wait long for my idea to strike a chord. The late Lee Romney, a prominent Fund supporter, immediately offered to build a chimpanzee house. Equally immediately I called Dr. Lemmon and asked him to come over to see Black Beauty and okay it and also to go over some ideas I had for the house. He agreed to do this, and it was not long before he also agreed to our having Nim. The kind of chimpanzee house we settled on was a large one which would have a whole gymnasium inside, a large porch outside and, at the rooftop, a captain's walk from which Nim and his future "significant other" could view most of the other animals that were around and feel themselves part of the whole operation.

While we waited for Nim's house to be built, I devoted a good deal of time to finding out everything I could about Nim's early life. It was not a difficult task. There were adult books about Nim, picture books about Nim, and even children's books about Nim. There were also many television shows and videos about him. One of the first things I learned was that he owed what he had become almost entirely to one man.

His name was Dr. Herbert Terrace, a man who had started his career as a graduate student of Dr. B. F. Skinner, a notoriously controversial experimenter, at least from the point of view of the animal community. Dr. Terrace had apparently been for some twelve years experimenting on white rats and pigeons. He had long been interested in chimpanzee language experiments, however, and had decided upon adding a chimpanzee to his repertoire. He had also decided that the best way for a chimpanzee to learn the human language was to have him brought up in a human family—as exactly as possible like a human baby. In other words he would sleep in a crib, dress in human clothes, sit in a high chair at meals, and be toilet trained.

Although he was at Columbia University and not the University of Oklahoma, Dr. Terrace was a friend of Dr. Lemmon's and persuaded Dr. Lemmon to let him have Dr. Lemmon's next newborn baby chimp. To this Dr. Lemmon agreed, provided only that Dr. Terrace agree to return the chimpanzee to him when Dr. Terrace's experiments were over. Dr. Terrace's next job was to find the right family to bring up his chimp baby. For this he chose Stephanie LaFarge, one of his graduate students who had just married into one of New England's most literary and artistic families.

When Nim was born on November 21, 1973, Dr. Terrace was away at the time but in any case decided to name him after one of his language-behavioral friends, Noam Chomsky, but was mercifully persuaded by another friend to simplify the name to Nim Chimpsky. In time the Chimpsky was also mercifully dropped. Dr. Terrace also persuaded Mrs. LaFarge, who was living with her husband in New York, to make the trip to the

University of Oklahoma to get Nim. She herself perfectly recalls everything about the experience in a moving essay which she wrote for me—one which she called "From One Mother to Another":

On the day I arrived in Norman, Oklahoma, Dr. Lemmon took me to see Nim's mother, Carolyn, and her new baby. I was a stranger in the chimpanzee compound so they greeted me with noisy agitation. Carolyn's behavior was different; she remained still but very watchful. She held her three-day-old infant close to her breast. I admired the way she shaped her body to hide the tiny creature, positioning herself between the baby and the outside world. "She knows," Dr. Lemmon said. He told me this was Carolyn's tenth infant. All the others had been taken from her within the first month. Dr. Lemmon said we would have to render her unconscious to "free the infant." We planned the transfer for the next day.

In bed that night, I dreaded the impending separation of this mother and her baby. She was losing her babies, not because she lacked the skills or motivation to be a good mother, but because there were human families throughout the United States eager for the privilege of raising a chimp in their home. And there were also Federally funded scientists eager to study the outcome of this "experiment" in interspecies communication.

The next day Dr. Lemmon and I reentered the compound. Without pausing, he fired the tranquilizing dart from his gun. Carolyn was startled but did not try to flee. She held tightly to Nim as the sedative took effect, and her body slowly tipped sideways. Dr. Lemmon rushed into the cage and grabbed Nim before the weight of her paralyzed form crushed him. Nim was screaming as he was handed to me.

When Nim was transferred from Carolyn's weakened grasp to my arms, I felt as if I was straddling two worlds. One was Carolyn's world in which this infant had a safe and predictable place. The other was my human world where his trajectory was unknown. My heart was pounding as I cradled Nim in my arms. I hoped he would feel comforted by this new heartbeat. Perhaps, I thought, it was not too different from the one he had known in the womb.

We crossed the threshold back into Dr. Lemmon's living room. Nim was simultaneously protesting and clinging to me. Dr. Lemmon's wife had laid out a baby blanket and a tiny diaper, but I couldn't detach Nim from his grasp on my neck and hair, so that is how he remained, more or less continuously, for the next three days. I learned to change his diaper without putting him down, and within the first few hours of separation from his mother's breast milk, he adjusted to for-

mula from a baby bottle. Eventually, I learned that when Nim was asleep, he could be wrapped . . . almost swaddled . . . in a blanket and then tucked securely in the twelve-inch vinyl soft-sided traveling case that I had brought for the flight to New York.

When I made a reservation, I had gambled on not identifying this infant as "nonhuman" to the airline. In first class, the stewardess was appropriately solicitous but too busy to pay close attention. The flight was well under way before she had time to ask my baby's name and to offer to warm a bottle. I guessed she was somewhat skeptical of my mothering style since I kept his head covered all the time. In truth, I wanted so much to part the blanket and show off this infant as I had done with my other three children when they were babies. Instead, I pretended to be asleep until close to landing. "Your baby is so good," the stewardess exclaimed. This time I let Nim's tiny hand with his long "aristocratic" fingers and hairy arm emerge from the blanket. Her shock turned to delight as she scooped up this passenger to show to the pilots.

Before leaving the Primate Center, I had gone to say good-bye to Carolyn. As I silently promised her that I would "do right" by her baby, I will never forget the way she looked at me. At the time, I was so excited by the challenge of caring for Nim that I was blind to the implications of what I was doing to these two sentient beings.

The family into which Nim arrived was considerably more than a family. It was, rather, a tribe. It consisted of Stephanie and three children from her previous marriage—fifteen-year-old Heather, fourteen-year-old Jennie, and eleven-year-old Josh—as well as Mr. LaFarge and three of his children, from his previous marriage, not to mention a schoolteacher who was a friend of the family.

One day Stephanie confided in Dr. Terrace that she had wanted to integrate her new family and her new home by having a baby. But with the coming of Nim she had agreed to try to satisfy this urge by having a chimpanzee instead. When her husband met her at the airport, however, Nim—even though just a few days old—had ideas of his own as to just how he wanted to be integrated, and when Stephanie's husband tried to hold him, he instantly and loudly demanded to be handed back to Stephanie. Some months later, when Stephanie and her husband were taking a late-afternoon nap on their large water bed and Nim was lying between them, Stephanie's husband reached out and put his arm around her—to which Nim's immediate reaction was to stand up and bite Mr. LaFarge. He did not succeed, as in Dr. Lemmon's case with his chimpanzee, in pushing Mr. LaFarge out of bed—but this may well have been either because, as a baby, he did not have the strength, or because it would have taken a large push to dislodge someone from a water bed.

From the very beginning, indeed from the very first day, Nim was spoken to only in sign language. Not only did all the LaFarge family know the language, so did a whole procession of regular researchers who were continually recruited by Dr. Terrace from Columbia and Barnard. Actually, to say Nim was started on being taught at a few days old was not accurate. He was literally forced to learn sign language on the very first day he arrived at Stephanie's house. In any case, from that first day on it is doubtful if any animal in history—and certainly never one that young—was ever subjected to such an ordeal. What this amounted to was that day after day, for almost four years, with only the briefest respites, Nim could not see anything, have anything, eat anything, play with anything, or even point to anything, without having it endlessly taken and turned, over and over again, into sign language. That he was not driven crazy is certainly remarkable.

Before Nim was even given a bottle he would first have to sign for the word "drink" by extending the thumb from the fist which was then touched to the mouth. From the time when he was two months old his teachers began to teach him the signs by physically molding his hands. At first Nim resisted having his hands molded, but he was soon forced to allow his teachers to mold his hands to make the signs for such words as *drink, up, sweet, give,* and *more.* Nim was particularly enamored of the sign for *more,* especially when it came after eating, drinking, being tickled, or having a pillow fight—the parts of his lessons he liked best.

Among Dr. Terrace's stringent requirements for all Nim's teachers was that no sign could be considered learned until three independent observers had reported its spontaneous occurrence, and even then the sign had to occur not only spontaneously but also appropriately and on five successive days. Besides all this sign learning, Nim at only three months of age had begun picture learning from magazines and books and before much longer was even painting pictures himself.

After Stephanie's home, Nim's chief "classroom" became a four-room suite at Columbia University. The room used as his main one—although in reality every room was a classroom—was a bare room just eight feet square. The idea of this was that the room was so small Nim could not very well romp around in it and would more or less have to concentrate on whatever the teacher was having him do. One wall of the room had a large one-way window on it which allowed whichever of Nim's other teachers were present to watch what the teacher of the moment was doing.

Under the window was a portal which contained a veritable studio full of video equipment, photographing every step of his learning. Working at this time not primarily with Stephanie but with a new and far stricter head teacher, as well as a large number of new assistants, Nim was expected to enter his tiny classroom each morning to hang his hat, sweater, and coat on a hook on the wall. At breakfast he was not only expected to sit in a high chair and eat with a spoon but also, when he finished, to wipe his face and hands as well as the chair. On top of it all, Nim was also expected to give his entire attention to his teachers and to ignore all other sounds that seemed to clamor for his attention. Some of these were the high-pitched squeal of rats which came from the psychology department's vivarium behind one of his walls. "The rats," Dr. Terrace mentions of the incident, matter-of-factly, "often did this when they were being taken out of their cages." On one occasion, despite all the locks that had been tested on his doors, Nim broke away from his teachers, entered the "Rat Room" himself, and proceeded to liberate all the rats and lead his teachers on a merry chase throughout a sizable part of the University.

One sharp change after Stephanie's reign as chief teacher came when another new teacher decided to teach Nim sign language in three stages. The first stage she called "Reception," which was completed when Nim did something to indicate he understood the sign. In other words, if the sign was "eat," the teacher would have to wait until Nim grabbed something and pretended to eat it or else pointed to some food in a picture book. The second stage was "Production," which meant that Nim was required not just to allow his hand to be molded by the teacher into the appropriate sign but to do the molding himself. The third stage was "Expression." In this

Nim had to both make a sign and use it in an appropriate context as, for example, making the sign for "drink" if he wanted something to drink.

The idea of all this was that at least one of the teachers believed it was better for Nim to learn, for example, the sign "cat" by sign language than it was to look at a cat in a picture book or even look at a real cat. Certainly this could be disputed. The sign for "cat" in sign language, for example, is made by putting both index fingers beside the nose and then pulling the fingers apart and rubbing them over imaginary whiskers. To say that it is easier to learn "cat" by that sign than it is to be shown a picture of a cat or better still a real cat is open, at the very least, to question. It certainly was in the case of Nim. In one of the very few instances when he was allowed not only to be with but also to play with a real cat he almost went crazy with the fun and joy of having the cat with him, and from that moment on the word and sign and everything else about cats was completely in his mind. To say it would have been better for him to go monkeying around with two index fingers and imaginary whiskers is simple nonsense.

With some of the teachers teaching one way and some

with totally different methods, all having to be learned by Nim, it was slow going—and particularly as Nim grew larger and stronger, there were increasing disciplinary problems. In fact, Dr. Terrace admitted that of all the problems he found with Nim's teachers, the question of how Nim should be disciplined for antisocial behavior, such as screaming and having tantrums and biting, created more controversy and emotion than any other. Nim's "Reception," "Production," and "Expression" teacher, for one, proposed that Nim be placed in a small, dark box called a "Time-Out Box" for bad behavior.

Other teachers, however, objected to this on various grounds, not the least of which being that as Nim grew larger and stronger it was more and more difficult to catch him and put him anywhere, let alone in such a box. One teacher suggested that every time Nim bit, the teacher should bite him back, but this too became unworkable because from the very first time it was tried Nim thought it was the best game ever. His teachers obviously enjoyed it considerably less, particularly since a chimpanzee has a full set of canines at ten months of age and a bite from such a set may have been enjoyable for Nim to deliver but it was hardly a game for the recipient.

Dr. Terrace then suggested that signs be used for Nim's disciplinary problems—that, for example, when Nim bit, the teacher should sign, "You bad. Me angry you." Or even, "If you bite, I leave."

While this seemed at first a good idea it too proved difficult because among other things, besides signs, Nim had learned how easy it was for him to get rid of teachers he didn't like by either screaming at them or, if necessary, biting them. Finally, Dr. Terrace was reduced to suggesting that some form of corporal punishment be used and suggested a kind of cattle prod, which was used at many primate centers, be tried. This method, while adjudged an improvement on the "Time-Out Box," was abruptly vetoed by a literal-minded teacher who equated Dr. Terrace's suggestion with electric shock therapy.

All in all, disciplinary failures or not, and by one method or another, and one teacher or another, Nim ended up comprehending some two hundred signs, for example: crayon, angry, green, hug, music, and hurt. He also learned to make such two-, three-, and four-sign combinations as: banana eat Nim, tickle me Nim play, more tickle, hug Nim, and sweet Nim sweet. During one two-year period Nim's teachers recorded him making some twenty thousand combinations of two or more signs. As if this were not enough, there were even instances when teachers recorded that Nim had made as many as eighteen sign combinations.

But still it was not enough. After Nim had worked for almost four years with his sixty teachers, Dr. Terrace summarily ended what he called "Project Nim" and returned Nim to Dr. Lemmon at the Institute for Primate Studies at the University of Oklahoma. Dr. Terrace had hoped Nim could learn to connect the signs he had so laboriously learned into grammatical sentences. He had also hoped that Nim would himself grow more and more interested in the project. Neither of these two hopes came about. In Dr. Terrace's opinion, Nim had not connected signs to make grammatical sentences and he had not grown more interested in the project. In fact, by the end, he had grown considerably less interested. Assessing these two findings, Dr. Terrace eventually declared that the project had been a failure and even added that no chimpanzees were capable of learning human language.

Certainly such endless examples as "Drink me Nim" and "Eat grape Nim" would seem perhaps to bear the doctor out since they were not clearly grammatical. But Dr. Terrace's giving up on the whole project and declaring it wholly a failure seemed to many people premature. Nim had been, after all, still a very young chimpanzee when "Project Nim" began—in fact, he had been just a few-days-old chimpanzee. And not only had he, in four years, been able to learn over two hundred signs, but also what is more important he had been able to connect as many as eighteen signs together—this achievement being what Dr. Terrace thought of as the key to learning human language, i.e., not just learning human signs but connecting them in a human grammatical sentence. If Nim had not actually learned to speak human language, he had at least gone as far down the road toward it as any other "signing" chimpanzee before him or after.

Stephanie LaFarge, closest of all Nim's teachers to him and indeed Nim's surrogate mother, if he had ever had such a thing, believes "Project Nim" was fatally flawed because, while it made Nim half human, it did not make his teachers half chimpanzee. In a letter to me she expounded her theory as follows:

I believe that Herb Terrace's belief in the B. F. Skinner behavioral system ignored what was really unique about Nim. In an animal as complicated as Nim, you really couldn't shape him into making him like the human being Dr. Terrace apparently wanted. At a certain point he just became bored and willful and just resisted and refused any more.

If I had been Dr. Terrace, I would have tried to create a very different environment from the one he tried to create. It could have been half human, all right, but it also should have been half chimpanzee. Half the time he would have put on the clothes and tried to be human, but the other half of the time we would have had to try to be in his environment—climbing trees and all the rest of it. Then the whole language thing could have come about naturally.

Dr. Terrace always wanted to know what was in Nim's mind. If he wanted us really to learn that and know that, the whole program should have been devised just as much to make us talk to him as it was to make him talk to us.

Let me give you an example. One day, when I was teaching Nim in his classroom at Columbia, a whole group of white rats got loose from the next room—he hadn't let them go this time, but someone had, and they came into our room. For the first time, Nim and I were on equal ground. We were both frightened, and it was not just teacher versus student, it was human and chimpanzee versus rat. Although it didn't last very long, and people from the lab came in and cornered the rats and took them away, what Nim and I had together in that situation had nothing to do with teacher/student, it had to do with two equal creatures together.

Let me give you another example. One day, when I was teaching Nim, my purse, which I had carelessly left open, fell off my lap. Before I could retrieve it Nim, who is incredibly quick, got into it and immediately proceeded to grab the few dollars which I had in there. When I tried to get them back, for the first and only time for all my relationship with Nim, he bit me. He didn't bite me hard, and the bite wasn't the important point. The important point was that he was so incredibly bright that he had seen his teachers and other people giving each other dollars and taking dollars from each other and he was convinced there was something very important about them, and although he couldn't understand what this was, since it seemed so important to them, then it must be important to him, too. Again, like the time with the white rats, although a very different time, we had a real relationship and, again, on a much more important level than just teacher and student.

Mrs. LaFarge left me with what I felt was a wonderful postscript about Nim. She told me that, one day, she took Nim out walking when they were at the Delafield House at Columbia where Nim was staying. Everybody had been very concerned that Nim didn't get a chance just to play in the grass, so Mrs. LaFarge made a particular effort to take him out in a large field. "Do you know what?" she told me. "Nim just hated it—he hated grass. He couldn't wait to get away from it and go climb a tree." I have often thought of this when people see Nim's house and complain that he doesn't have enough grass. I do not tell them he doesn't like grass, but I often want to.

As we waited for Nim's house to be completed and contemplated the extraordinary half-human half-chimpanzee who would shortly be coming to us, all of us felt one thing strongly—this was that whatever life Nim had had in the past, from now on he would never have to learn a single sign unless he wanted to do so. Most of us had picked up many signs, enough, we felt, so that we could give them to Nim if he so desired.

But from the time he arrived we talked to him in the King's English.

We talked a good deal of the time, too, of course, to Sally, the "significant other" chosen by Dr. Lemmon, and we quickly learned that as extraordinary as Nim was in his way, Sally was in hers, too. An ex–circus chimpanzee, much smaller than Nim but with double his charm, she turned out to be the perfect companion for him after we went through some difficulties at first. One of these was that Sally had much better manners, both table and otherwise, than Nim. Indeed, in their first few days Nim would invariably gobble his own food, then push Sally out of the way and gobble some of hers. And when we would offer them both a treat he would invariably eat hers first. Etiquette among animals at the Ranch was very important to us, and we immediately went to work on this breach of it. What we eventually devised was that, when it came to food, we not only gave it first to Sally, but also we gave it in such a way that if Nim grabbed it he did not get any food of his own. Soon we had Nim waiting for Sally to eat before he would touch so much as a morsel—just like a properly brought-up bridegroom.

The fact that Sally had no knowledge of signs was not

important. She soon picked up from Nim if not signs, then the same kind of thinking he did. We would ask Nim, for example, if he thought it was going to rain, and immediately he would look up and shake his head, exactly as a person might do. I remember, for another example, the first time I brought Nim his toothbrush and toothpaste—he not only liked brushing his teeth, he was positively crazy about the taste of toothpaste. Still, when I brought the brush and paste to him, he gave me an unforgettably patronizing look and pointed toward where the brush and paste were kept. As I went over there I saw what he wanted and I had forgotten—his mirror. But that patronizing look was one I didn't easily forget. He didn't even need to speak his feelings, it was so obvious he was saying, "How on earth do you expect me to brush my teeth without a mirror?"

From the beginning both Nim and Sally liked magazines and picture books and also painting pictures. We have a television set hooked up for them, too, but they spend very little time watching it and even, to my irritation, ignored my peerless critiques in *TV Guide*. Nim was, however, extremely fond of various new sports we rigged up in the gymnasium for both him and Sally. His favorite single apparatus was a tire, which he liked to roll around and which he also liked to roll out on the porch. He would also sit on the tire as he passed the time of day with you. The tire, incidentally, showed how incredibly strong he was. He could, for example, hold a pumped-up automobile tire at arms' length, with both arms extended, and actually compress it.

What he was doing is worth trying sometime—holding something at arms' length, with both arms extended, and seeing how much strength you have. You will not be surprised when you are told, as we were, that a full-grown chimpanzee has the strength in his arms of eight men.

Adding to his gymnasium, Marian Probst went to the trouble of getting Nim a full basketball board and basket, as well as a basketball. For a time it occupied both Nim and Sally very well, but failed completely in the slam dunk. Nim's preference for this special feature of basketball was to jump up on the hoop, hold on to the rim of the basket, and then do such a mighty dunk that the force of his weight brought down not only the ball and the basket but the whole backboard.

In time, everybody who worked at Black Beauty became as fond of Nim and Sally as any animals at the

Ranch. As for the tour groups, they couldn't get enough of him and spent as much time as possible with both him and Sally. To many people who knew about Nim he was a major animal celebrity, and not a few people came from all over the country, and in many cases from abroad, to visit him. Nim liked the attention but took it very casually—after all, he had had it since he was only a few days old—and often when some chimpanzee researcher would come and study him for a long time, Nim would get bored with the whole thing and jump down from his tire or wherever he was and start a war with the Ranch dogs—which was one of his favorite methods of exercise. First he would hoot and holler, and roll his tire back and forth across the porch, and then when the dogs came he'd up the game to riotous levels. Sometimes indeed the situation would get so totally out of hand that Nim would have to be quieted by Chris, who would placate him with some food or a different game.

Often Nim would declare war not just on visitors or dogs but even on nature itself, and particularly on thunderstorms, which he detested. If, for instance, a thunderstorm came on in the middle of the night he was eminently capable of waking the whole Ranch with his noisy wrath until it was hard to tell which was the noisier, Nim or the thunderstorm. There were even occasions when, as carefully built as Nim's house was—all the way from the gymnasium to the captain's walk front portage—Nim would somehow manage to get out. For me, the most difficult of these occasions was when Tom Regan, a friend of mine and leading animal advocate, came to do a television show about the Ranch in general but Nim in particular. At the conclusion of a show, he wanted me to sit outside with Nim and talk about how the Ranch came about. I explained to Tom and his crew that I couldn't do this with Nim outside, but then, very foolishly, I suggested I *could* do it with Sally.

Of all the dumb ideas I've had about our animals, this surely ranked as one of my dumbest. Hardly were Sally and I in our appointed chairs and the television cameras rolling than there was a hoot and a holler, a rush and a blur, and who should blow by, heading at full tilt for the main house, but an extremely annoyed Nim. How he had managed to get out we never did discover, but Tom and his crew promptly called the show a wrap and hightailed it after Nim, totally deaf to my loud pleas that they should try to go anywhere except back in the house. I well knew that wherever else Nim was determined to go, the house would be his first port of call—obviously because he had been brought up in a house, and he knew there were all sorts of goodies residing there.

At this juncture Sally, too, decided that joining Nim was far more fun than continuing to sit there in front of nonrolling cameras, and without so much as a passing thought about the idea that the show must go on, she too careened after Nim and into the house. Most of the crew made for the living room, which was unfortunate because it was right next to the kitchen. By the time anyone got anywhere Nim had not only gotten into the refrigerator but had managed a massive sacking of same. As for myself, on my way to join my routed troops I went through the bathroom and grabbed at Sally. My idea was to take her prisoner in the bathroom and at least make an attempt to divide and conquer our foes. Although Sally was normally the gentlest of chimpanzees, she was now so much under the influence of Nim's anger that my attempt to capture her came to naught.

The very next thing I knew, however, someone was knocking on the bathroom door and I immediately figured out, in my inimitable way, that it was one of Tom's crew looking for a safe haven. I reached to open the door when the whole door was literally wrenched off its hinges and I was pushed over backwards, down into the shower bath. There, standing over me, were Nim and Sally, obviously highly pleased with the carnage they had wrought so far. They were now clearly deciding, by looking at me, what fun it would be to add to their policy of taking no prisoners, by making no exception in the case of the boss.

The backward position into which I had been so unflatteringly shoved left little hope that I could assume a commanding demeanor, let alone try to thwart whatever plans they had for me. I even considered connecting about eighteen "No, Nim, No" signs. I gave this up, however, and just looked up at both of them, one by one and then, addressing Nim in particular, I scolded him, as firmly in that particular position, as I could. Nim, I said, you have terrified every single soul in this house. Please, if you can't be good, at least make an effort to call a moratorium.

I honestly believe Nim understood what I was trying to say, even if the word "moratorium" might have baffled him. In any case, he put his head down as he often did when he felt ashamed and, still holding onto Sally, he left the bathroom and went down toward the bedroom. His change of direction gave me time to shoo the pathetic remnants of my army out of the house, with instructions not to come back until the coast, including both Sally and Nim, was clear. In time, our then-manager of the Ranch appeared with his tranquilizer gun—something which Nim knew exactly what it could do to him and, after one look at it, and still with Sally in tow, he went docilely back to his own house.

Long after the experience was over, however, I reflected upon how, sitting back there in that bathroom shower, I had been entirely at Nim's mercy and I thought long and hard and gratefully that, angry as he was, he had not tried to hurt me. I also reflected upon how our half-and-half child was clearly still half chimpanzee but he was also, at least where humans were concerned, more than half human.

AS THIS BOOK WENT TO PRESS, SALLY UNFORTUNATELY passed away. Her obituary was printed in many animal magazines and posted on the Internet at the Fund for

Animals' Web site (http://www.fund.org). The obituary read as follows:

SALLY JONES, 1950–1997

Sally Jones, approximately 47, longtime resident of the Fund for Animals' Black Beauty Ranch in Texas, passed away at 4:00 P.M., March 21, 1997. As an infant chimpanzee, Sally was captured in Africa around 1950. Like many wild-caught chimps of this era, she was shipped to the United States, and spent the next twenty years working in the circus. She walked upright, performed ballet, and was an accomplished bicycle rider, even though she still had buckshot in her abdomen from her capture. This injury kept her from reproducing.

In 1970, she was acquired by the Institute for Primate Studies in Norman, Oklahoma. She spent the next thirteen years in Oklahoma, living in the large chimpanzee group which was involved in behavioral and cognitive research at the University of Oklahoma. In 1982, following the sale of the Oklahoma chimp colony to medical research, she came to live at Black Beauty Ranch, and became the companion to Nim Chimpsky, who had also been at the Institute. The vast majority of Sally's companions at the Institute ended up at the Laboratory for Experimen-

tal Medicine and Surgery in Primates, affiliated with New York University.

In 1993, Sally suffered a stroke, from which she fully recovered. Around this time she was diagnosed with diabetes, which had been successfully managed over the last few years. Sally was one of the oldest living chimpanzees in the United States.

Sally is survived by her companion of fifteen years, Nim Chimpsky, and by Ranch Manager Chris Byrne and the staff of Black Beauty Ranch and countless friends, all of whom were touched by her sweet disposition and amiable character. She will be fondly missed by all of them. The new expansion of the chimp outdoor play area at Black Beauty Ranch will be dedicated in Sally's memory.

Nim was inconsolable about Sally. After she was gone he spent countless hours just staring at her bed and ate so little that everybody at the Ranch went into high gear to find another companion for him. Many animal societies helped us in the search and, in the end, we were presented with no fewer than three new creatures to take Sally's place—which, considering how wonderful Sally had been, was as it should have been. There are two females, Kitty and Lulu Belle, and one young male named Midge.

All three arrived separately, which made our plan to introduce them to Nim much easier and also enabled them, again arriving separately, to get used to one another.

Midge came first and, in a way, this was good because Nim could get used to the idea of another chimp before coping with two females. Twenty-two years old, a laboratory chimp from birth, Midge was so accustomed to his boxlike laboratory environment that until he came to us he had, we were told, never seen the light of day. Needless to say he was soon as little out of the light as possible and unpredictable, playful, and none-too-grown-up for a chimp his age. He has nonetheless proved both a fine companion and an almost ideal young new friend for Nim.

Kitty, thirty-five, also a laboratory chimp and known as a "breeder"—she had fourteen babies, including two sets of twins—was meek and somewhat withdrawn when she first came to us but has since formed a strong friendship with Lulu Belle and, something of a cocktail party hostess herself, has turned out to be an excellent mediator at forestalling little fracases. One of her favorite sports is crawling inside a barrel and playing her own game of "hide and seek" from all three other chimps. On one occasion a Ranch hand, thinking Kitty was outside and doing a first-rate job cleaning her enclosure, suddenly just before he left heard a strong sigh of contentment. Looking over at the direction it came from, he saw the edge of Kitty's head peeking out from her barrel. He tiptoed out of there.

Lulu Belle, thirty-three, is a bulky chimp, almost a miniature gorilla, who makes a practice of walking very straight on her hind legs. She takes no nonsense from either Midge or, remarkably enough, even Nim, and yet will immediately take Kitty's side in any problems with the males.

Altogether, Black Beauty Ranch has parted with a longtime friend, but it has made up for it to a large extent by making it possible for Nim to have a very special new family.

Chapter Five

OF BUFFALO, OF ELEPHANTS, AND OF PRAIRIE DOGS

CALL THEM BUFFALO OR BISON AS YOU choose—the words are interchangeable. But whatever you call them, you should never forget that almost throughout the nineteenth century they suffered the most monstrous animal massacre in American history. According to authorities, the most widely accepted total figure of the slaughter is sixty-three million one hundred fifteen thousand two hundred animals.

Ed Park, author of *The World of the Bison,* perhaps most memorably stated the difficulty of anyone even trying to contemplate such a figure. "Just how large a number," he wrote, "is 63,115,200?" He then answered his question himself, as follows:

Imagine a long, single-file column of buffalo, head to tail, walking past you, one every two seconds. Take a counter and start counting them—one every two seconds, hour after hour, day after day—without pause.

During the first minute 30 buffalo would file past; in the first hour 1,800 would go by. If you could stay awake for the first twenty-four hours, you would count 43,200 buffalo. The days would roll by and sometime during the twenty-fourth day the one millionth buffalo would hurry past.

Months would drag by, the seasons would change, and you would become tired of counting. Buffalo adding up to several millions would have passed before your eyes in seemingly endless file, with still more to come. At the end of the first year your counter would show a total of 15,768,000 buffalo. But the end would not be in sight!

The second year would pass, then the third, and finally the fourth, with even an extra day added for a leap year. As the fourth *year drew to a close, the last buffalo would hurry by, and you could look at your counter one last time! 63,115,200!*

Why were the buffalo massacred? It is a good question but unfortunately it has no even reasonably decent answer, mostly because they were shot for so many reasons. First of all they were shot because, like so much of "sport hunting," it was good fun, and the buffalo—placid, gregarious animals who liked to travel in herds and had notoriously poor eyesight, poor hearing, and only a moderate sense of danger—were easy prey. Second, they were shot for just about everything about them, for their hides, for their horns, and even just for their tongues—which, cut out of their mouths and roasted, were regarded as a delicacy and brought sizable amounts of money. Third, they were shot in contests, over and over again.

At first, in the early 1800s, the hunters' guns were not very effective but later guns, particularly the Sharpe "Buffalo Guns," which fired long slugs and were accurate as far as fifteen hundred feet, were cruelly effective. As Park and other writers point out, with such a weapon a hunter could kill over a hundred buffalo without moving from one spot. They used forked stakes at their "stands," because the rifles weighed over twenty pounds. And, with such a weapon the hunter could kill incredible numbers—250 in one day, 3,000 in one month, 5,700 in a year, and perhaps 30,000 in a lifetime. Particularly famous assassins, like Buffalo Bill and our old friend "Doc" Carver who, patient readers will recall, inaugurated the "diving horse" craze, both regularly lied about their deadly accomplishments but they were considerable.

Most of the illustrations and paintings of buffalo hunting in the nineteenth century show brave hunters on horseback riding right up to the side of a buffalo and shooting the animal dead. Stories about the hunting almost invariably emphasize what formidable foes buffalo were. Cy Martin, in *The Saga of the Buffalo,* for example, tells about going after bull buffalo. "With their stout horns," he claimed, "a bull could rip the rope-tough prairie sod to make himself a dust wallow, toss a whole wolf pack, disembowel a horse, and even carry a horse and rider aloft for a hundred yards before finally hurling them to the ground." At the same time, Martin also admitted, however, that the poor eyesight, poor sense of hearing, and even poor sense of smell of the buffalo made them about as difficult to shoot as cows in a barn.

Actually, at the height of the nineteenth-century hunting massacre, the hunters did shoot cows—buffalo cows. They did not even bother with the bulls at first; they purposely lung-shot the cows in the cruelest possible man-

KEITH MCELVAIN

Baby buffalo are red. Here, Hope and her mother.

KEITH MCELVAIN

But the baby's best protector is the bull, One-Eyed Jack.
This is as close as he would allow us to get.

MARY DA ROS

Tara, alone and without a friend for thirty-five years, entwines her trunk with Conga's welcome.

MARY DA ROS

Today, Tara is a curious combination of timidity and rambunctiousness.

MARY DA ROS

Conga and Tara—ablution time.

KEITH MCELVAIN

Chris with three special loves. Note Tara's trunk covering Babe's injured foot.

CHRIS BYRNE

Mary Da Ros and friends—one of them one of our many three-leggeds.

MARY DA ROS

Baby aoudads take to their bottles.

Kinkajou

Arctic fox

Wild pig

PAUL HOWELL

AMORY WINTHROP

Buffalo and burro share and share alike.

MARY DA ROS

Prairie dog secrets.

ner, so that they would bleed profusely. Meanwhile, as they stood or fell dying, the rest of the herd would stand near them as if to protect them, but of course could not and would also be killed.

Perhaps the favorite sport of all buffalo hunters was shooting them from trains. The trains not only panicked the buffalo into flight but also kept them on the train tracks while trainmen, who had buffalo guns prepared, would open the windows and, depending upon which side of the track the buffalo fled, let the passengers hold their guns on the windowsills and fire at will. The trains outran the buffalo, of course, but it was all great sport while it lasted, and even after it was over went on because, when most of the buffalo were either dead or dying, the trainmen would stop the train and the killers would be let out to cut off tongues. Most of the time they did not bother to make sure which buffalo were dead and which were just dying—they cut out their tongues anyway. There was such frenzy to get to these train-shoots that to get to one of them General George Armstrong Custer, as might be expected, got so excited he not only killed a buffalo but his horse as well. Calves were, of course, also victims. All in all, it was such unbelievable slaughter that one railway engineer said after a good train-shoot that it was possible to walk for a hundred miles along the Santa Fe Railroad by stepping from one buffalo carcass to another.

The most famous of all buffalo hunts was undoubtedly one that was put together in 1872 by the U.S. Government along with General Philip Sheridan and Buffalo Bill—all for the benefit of a distinguished visiting foreigner, Grand Duke Alexis, the son of the Czar of all the Russians. The Grand Duke arrived, as befitted his station, in his own private railroad car at Hays City, Kansas, where General Sheridan had thoughtfully provided him with a war dance staged by the Sioux Chief Spotted Tail.

After these festivities were over, Buffalo Bill not only gave His Majesty a lesson in shooting but also went so far as to see he was mounted on Bill's favorite mount, Buckskin Joe. They then rode off together not, unfortunately, into the sunset, but after their first buffalo. The Grand Duke got so taken with it all that he missed his first shot completely and would have also missed his second, except that, perhaps in the spirit of the war dance, he not only put two pistol shots into an animal but whooped several times in joy. Finally, Buffalo Bill chased the buffalo down a ravine and carefully arranged for the Duke to fire two more final shots at the wounded animal. While all this was being accomplished the Grand Duke, by now totally hysterical, dismounted and slashed off the animal's tail, whereupon he remounted, this time the dead buffalo, and shouted and yelled in Russian, all the while being dutifully toasted with champagne by his Russian attendants.

By the time of that hunt the whole matter of buffalo massacring was intrinsically linked on the part of the U.S. Government with primarily getting rid of, if not actually exterminating, as many Indians as possible. Not only were both U.S. President Grant and General Sheridan prime participants in the idea of getting rid of buffalo to get rid of the Indians, they also favored the idea of massacring Indians as relentlessly as they had wasted Union soldiery at Cold Harbor and the Wilderness and on other battlefields in the last days of the Civil War.

General Sherman, for example, whether influenced by the Grand Duke's visit or not, wrote a letter to General Sheridan in which he suggested that Sheridan stage an even bigger buffalo hunt—one that, he suggested, would make "a clean sweep of the Indians." General Sheridan did not follow Sherman's instructions, but President Grant's Secretary of the Interior, Columbus Delano, publicly stated that "the elimination of the Indian is impossible while the buffalo remains upon the plains." Meanwhile, General Sheridan, hearing that the Texas legislature was about to pass a vote asking for a moratorium on the slaughter of the buffalo, went himself to Austin and fought against the bill. "The buffalo hunters have done more than the U.S. Army to get rid of the Indians," he stated. "They should have a medal and should be sent powder and lead until the buffalo are exterminated."

The bill, of course, failed. As for President Grant, it was in his administration that a bill was introduced in Congress to "make it," it stated, "unlawful for anyone not an Indian to kill, wound, or in any way destroy any female buffalo of any age found at large within the territory of the United States." Even though the bill was passed in both the Senate and the House, not only did President Grant not sign it, instead, as one historian noted, "He placed the document in a pigeonhole for the India ink to become a rich brown before he touched the bill with his lamp and lighted a cigar with it."

E. Douglas Branch, in his book *The Hunting of the Buffalo,* recalls Texas cowboy Charles Norris's story of the last herd of buffalo he ever saw—which was in the Texas panhandle:

I came on in the May of 1886; I was driving a bunch of horses from Coldwater to Buffalo Springs, saw the buffalo herd about three miles off. I did not have my gun with me then, so on my return from Buffalo Springs I went to the camp, and waited: it was pitched near the only water in the region, and I knew that in a day or two the herd would come to the water. On the second day the whole herd appeared. Now I had a good chance to count them, and they were 186. A number of them amused themselves by jumping off a bluff into the water four feet below them, then running around to a low place to jump off again. As soon as we had seen all we wished, we fired.

As for the last large herd of northern buffalo, these belonged to a curious character by the name of Michell Pablo. Orphaned early in life, half Mexican and half Indian, he was raised by a Scottish trader who taught him how to raise buffalo—how to buy and sell them and, most important, how to defend them from hunters. By the early 1900s Pablo had enough buffalo in his herd to interest the Canadian government in buying them from him for the then-princely sum of $250 per animal. It was to be cash on the barrelhead when the buffalo were delivered to the railroad station in Ravalli, Montana.

Pablo, like so many other people whose main interest in the buffalo was making money from them, went to work at once. He hired cowboys and had them drive the buffalo cows and calves to the cattle cars he also had them build at the Ravalli station. The cowboys did their job well and drove the calves and cows all the way to Ravalli and even saw them move immediately up the ramp, into the cattle cars. Pablo had, however, made one mistake. He had done nothing about the bull buffalo. In any event, one of the bulls promptly corrected the mistake. Following the calves and cows, he trotted right into the car all right but, instead of stopping there he trotted right on by them, and, where the cattle car ended he trotted right on through that too and all the way out into the valley where he was faithfully followed not only by all the cows and calves but also the other bulls.

Although seventy years old, Pablo was a tall and rugged fellow, and no greenhorn when it came to round-upping. While critics laughed at his failure, by the next winter he and his cowboys had built a railroad corral that was not only far tougher but also one that did not need to be built all the way around because one whole side was sheer, unclimbable cliff. Once more his cowboys were importuned to drive the cows and calves safely in. And once more there was trouble. The same wily old bull who had led the first buffalo charge to freedom this time outdid himself and led the charge right up the supposedly unclimbable cliff and on again into the valley.

For his third attempt, Pablo decided on a different approach. He located a small herd of about one hundred buffalo and ordered his cowboys to cut out a third of them and run them into the railroad cars, but this time the old bull and his fellow bulls, who were by now beginning to enjoy the whole game thoroughly, ended the plan by starting their own stampede with the rest of the herd. Robert Froman best tells the story:

This time, finding a herd of 100 grazing together, the riders cut out 30 and headed them toward Ravalli. The others stampeded up the valley with a wild uproar of thundering hoofs and reverberating bellows. Hearing this, the smaller bunch turned, charged straight back into the line of riders and rejoined the main herd. Eventually the cowhands found a small group grazing alone and got them into the corral. But a big bull inserted his horns under a 2 x 12 plank, heaved, tossed it over his shoulder. Then he backed up for a short run, splintered a couple of other planks and led the bunch to freedom.

Pablo now decided to try loading only cows and calves. After repairing the corral, he led his men out after the buffalo once more. It proved the wildest fiasco so far. The chase lasted from early morning until dark, covered some 60 miles, and ran down not a single buffalo.

To Pablo's credit he kept on trying, and in his last two attempts he did indeed get some buffalo into the railroad cars to stay. But once again they were not bulls, they were cows and calves, and once again Pablo's effort was fraught with the same difficulty as before. In one whole year, 1908, he failed to load even one single buffalo even though by this time he was reduced to trying a brand-new method—first driving the buffalo down an incredibly elaborate V-shaped trap which led first to a river and then to a beach. Here they were supposed to be crated one at a time and hauled into the railroad cars but for the final time once more it was a total failure. The last hundred bulls not only battered their way out of the crates but broke everything in sight, from the V-shaped traps to the railroad cars.

By now Pablo, at seventy-six, having spent far more money than he had taken in, gave up. But he had one last string to his bow. In a final effort to get his last hundred bulls to Canada he issued an invitation not only to Canadian officials but also to Americans to take part in what he promised would be the "World's Largest Buffalo Hunt." Unfortunately for him by this time the entire Canadian public, as well as most Americans, had tired of Pablo's war and had gone so far as to be on the side of the buffalo. The Montana Game Commission, in one of the very few times on record that they had ever done anything to help the buffalo, canceled the hunt.

PABLO DID NOT LIVE TO SEE WHAT FINALLY HAPpened to his bulls, but his bulls triumphed in the end and became the first animal citizens of the country's first National Park—Yellowstone, which became the country's first Federally protected wildlife sanctuary. Unfortunately, before many years had passed, hunters, lured by Yellowstone's easy pickings, went in and shot buffalo anyway. When this happened, however, President Grover Cleveland, who, when it came to buffalo had more gumption than Grant, Sherman, and Sheridan put together, promptly called in the Army and not only ran every single hunter out of the Park but also saw to it they stayed run out. I was named for President Cleveland and I've always liked both him and being so but never was I more proud of him than on that occasion—although it should perhaps be mentioned that I was at the time, in age, minus twenty-eight.

In any case, a century later, by which time the buffalo had become primarily a meat animal—one advertised if not particularly well enjoyed as "beefalo"—the hunters would be back again in Yellowstone, and this time there would be no President Cleveland to stop them. These hunters were, of all people, Department of Livestock agents and they were shooting the buffalo not just on behalf of Yellowstone Park but also on behalf of the sovereign State of Montana—a state that, in 1985, to its everlasting shame, passed a law authorizing the hunting of any buffalo who stepped over either the northern or western boundries of the Park.

A hundred years of persecution had not been enough. Now the Park Service, the State of Montana, and even the Department of Agriculture and its hired flunkies, the Department of Livestock, had turned on the buffalo. It was all over the fact that all of them had apparently been terrified at birth by a disease called brucellosis—one that the aforesaid Park Service, the State of Montana, the Department of Agriculture, and the Department of Livestock all believed would cause their beloved cattle to abort.

The fact that the Park Service was not supposed to be in the business of cattle raising to begin with, and the fact that Montana was, in comparison to almost any other state you could choose, a relatively poor cattle state, was either not noticed at all or, if it was, not given any consideration. The Department of Agriculture had stated that Montana was a "brucellosis-free state," and that was that—and this despite the fact that there has never been a documented case of brucellosis being transferred from wild buffalo to domestic cattle.

Nor had anyone even noticed or given any consideration to the fact that there were more than ten times as many elk in Montana as there were buffalo, and that these elk moved back and forth between Yellowstone and Montana and again nobody noticed, took into consideration, or even said a word about this. It was all very well to kill buffalo for carrying a disease to other animals they didn't carry it to, but it was not all right to kill elk for the same reason. Elk, after all, were not only mother's milk to every gun-loving Montanan from one end of the state to the other, they were also, considering the hunters, the outfitters, and even the tourists, one of Montana's major industries. Because of this the elk were certainly not an animal that any Montanan, with the possible exception of a card-carrying Communist, would want to interfere with for the sake of a little disease called brucellosis.

So, indeed, Montana reasons, or at least tries to—which for them is not always easy. As many of our patient early-chapter readers may remember, we were not entirely generous, when it came to animals, in our assessment of the State of Texas in that regard. But let no reader make a mistake. Compared with the State of Montana, when it comes to animals the State of Texas is close to the Promised Land. For one thing, the whole workings of Montana's politics, from the Governor to the Senators to the State Senators to their one Congressman, is totally antianimal or at least antianimal having to do with anything except using them, eating them, or shooting them. At one time, at a hearing in Washington, I was queried about the buffalo by Montana's then-Congressman, Ron Marlenee. "Mr. Amory," he asked me, "is it true that you said Monantans would shoot their mother if she was on four legs? I would like to know if you include my mother in that assessment." I told him I had not so far but I would be happy to do so if he had felt offended by my having left her out.

Not only Montana's politicians but also their judges are, if not totally antianimal, at least nearly enough so that the distance between their not being totally so seems infinitesimal. The Fund for Animals, for example, has been before Montana Judge Charles Lovell in one lawsuit or another—all over the buffalo—as far back as I can remember. Yet in not one of these suits have we ever had, in my estimation, remotely fair treatment, let alone fair judgment. Again, I did not know Judge Lovell's mother, but if I had I would not have hesitated to warn her about going out of doors with her son in hunting season.

In any case, with Montana's law about shooting the buffalo outside of the Park, the animals were caught in the middle between the Park Service and the State of Montana over the matter of brucellosis. And, although Montana was of the buffalo's adversaries clearly the most unreasonable of the two, the Park Service was not that far behind them. We had learned in the Grand Canyon fight, over the burros, that not only is the Park Service not a service—at least not a service to animals—it is actually against doing almost anything for animals. The Service relies, instead, on what they call "pristine management." For many years I have done my best to figure out what they mean by that, but I never have. All I know is they do nothing to help animals—when they are hit by cars or snowmobiles, when they are dying or drowning or whatever—the Service does not lift so much as a finger, ever, to relieve them from any pain or misery of any sort. Indeed, instead, the Service prides itself on saying they are trying to keep the Park as near as possible to what it was when the Indians were there. About this I can only say that I have long fervently wished they would go all the way with that idea and get themselves out of the Park and let all the Indians come back.

Texas people, as we have noted, are often inclined to be on the eccentric side. In comparison with Montana people, however, they are close to being straight arrows—and this includes considering both states' wealth of multimillionaires. Not since King Ranch heyday days, indeed, has any Texan held a candle in weirdness to Montana's Ted Turner—at least when it comes to ranch buying. So far Mr. Turner owns five ranches in Montana, three in New Mexico, and one in Nebraska—for a total acreage of 1,305,000. On these ranches are some 12,300 buffalo which Mr. Turner raises for slaughter for his abiding belief in "beefalo" as opposed to cattle beef. As for the other animals on his ranches they are fairly regularly assassinated by Mr. Turner and his family and his guests, although he allows nonguest hunters to join in the assassinations for a price of $9,500 per five-day period.

Mr. Turner's pride and joy among his ranches is a log-cabin "hideaway," as he calls it, which he had built for Jane Fonda and himself. In this "hideaway" there are not only all manner of animal skins and even carved chairs with bear claw handles, there is also, as at a majority of Turner ranches, an extraordinary number of mounted heads. More than one visitor has noted that, as Mr.

Turner increased in age, the mounted heads, even of different animals, seemed to bear more and more resemblance to their owner.

Second only to Mr. Turner as an odd Montana rancher was the late magazine mogul Malcolm Forbes—balloonist, motorcyclist, and peripatetic squire of Elizabeth Taylor. Although his ranch is not as elaborate as Mr. Turner's digs, Mr. Forbes did his best to make up for this by calling it the Royal Teton. Even this, however, soon paled for him, and he decided to put the entire place up for sale. Much to the distaste of many Montanans in the vacinity, the purchaser turned out to be, even by weirdo Montana standards, something special. It was called Church Universal & Triumphant—one with which we came into close contact when we were looking for an area that could be fenced and would have at least given the buffalo some place to go where they would not be shot. What we hoped Church Universal & Triumphant would see their way to doing was to let us build such a fence on their property. Unfortunately, Church would not even consider doing this until we understood more about their Church and what it stood for.

Dutifully, we undertook to do this. The founder of

the group, we learned, was a woman named Elizabeth Clare Prophet, and her method of teaching salvation to her followers was by the teachings of what she called the "Ascended Masters"—of which she was one but which also included Jesus, Buddha, Mohammed, St. Germain, the archangel Michael, the angel Gabriel, Merlin the Magician, and several others with whom we were not familiar, including El Morya, Hilarion, Ray-o-Light, and K17. Mrs. Prophet, we were happy to learn, was not the only woman among these "Ascendeds"—so were, we found out, the Virgin Mary and Mary Magdalene. She also regarded herself as the reincarnation of Guinevere, King Arthur's wife, and Marie Antoinette.

Born in Red Bank, New Jersey, Mrs. Prophet had been on a rooftop in Boston when she felt the urge to call out to St. Germain to come and get her because they were, as she put it, "twin flames" and were destined to be married. Although St. Germain did not come and marry her, she did marry three other men, the previous one of whom was named Prophet. The one on whom we pinned our hopes for getting our buffalo fence, however, was her then-present husband, by the name of Ed Francis.

Mr. Francis told us that Mrs. Prophet-Francis had

started Church Universal & Triumphant in California but had been directed by St. Germain to go to Yellowstone because, St. Germain told her, Yellowstone was on sacred ground and because, as he also told her, spiritual life required mountainous terrain. Mrs. Prophet-Francis was clearly no ordinary person. She did not like jewelry, we were told, but St. Germain had told her not only to wear it but to wear rings on all ten of her fingers. If Mrs. Prophet-Francis was expansive about jewelry, however, she was far from so on other questions—particularly on sex among her followers. In fact, she limited contact among single men and women among her followers to no more than five minutes of unsupervised visits, and married people were told they could have sex only twice a week, and then only if they had "decreed" for twenty minutes both before and after they had had it. It—the "decreeing," not the sex—was apparently very important in Church Universal. What it involved was repetitive chanting, over and over again, depending upon how much "bad karma," or sin, or sex, or both, one had the misfortune to take in.

At first those of us, including myself, who had contact

with Church Universal found it very difficult to get through all the ascending and the decreeing, at least to get far enough through to get somebody to decree a fence. Finally, though, we prevailed upon Mr. Francis to do so and to agree to pay for half of it. For two years the fence worked fine until St. Germain showed up again and apparently told Mrs. Prophet-Francis that a twelve-year "Cycle of Darkness" was about to fall on Montana. On hearing this, she instructed her followers to drop everything else, forget about our fence, and build underground shelters and stock them with food and guns. Even our friend Mr. Francis apparently got into the spirit of the thing and went out and procured so many weapons he soon found himself in jail. At this turn of events Mrs. Prophet-Francis divorced him and got another husband, by which time our whole plan for the fence was down the drain.

The Fund, however, had other irons in the fire. Our wildlife expert, D. J. Schubert, and I devised a plan to stop the buffalo from getting out of the Park simply by hazing them back in again. It was remarkable how simply the plan worked. Several times Schubert and I did it together, and more than once Schubert did it all by himself. He taught me, among other things, how to talk to the buffalo as we were doing it.

The Park Service, of course, thought nothing of our hazing. At first I was convinced the main reason they were against it was because they were afraid of the buffalo. But even this was not the main reason for it, because they had helicopters and could have done it all easily from the air. The real reason was that the Park Service, the State of Montana, the Department of Livestock, and the Department of Agriculture did not really want any solution to the buffalo killing. They just wanted the killing to go on.

In the end, horribly enough, they got their way. The first idea for the killing was a lottery in which hunters would draw a number to get to kill a buffalo. This hunt was truly a murderous affair. One youngster who could not have been more than fifteen years old shot at "his" buffalo over and over again. The kill itself took more than twenty minutes. The Park Service and the State of Montana eventually gave up their lottery and instead opened up the hunting to anybody. This too was equally cruel and murderous. One buffalo, shot over and over again, tried to get to his feet a total of forty times before he was finally killed.

For their third effort the Park Service and the State of Montana gave up lottery hunting and even regular hunting and instead used state agents to do the job. But even with the state agents from the Department of Livestock there was little improvement, at least as far as the cruelty was concerned. Soon there was clear evidence that the buffalo were not just being killed because they were getting out of the Park; they were actually lured out of the Park for the killing. Depending upon how deep the snow was on the ground, and how little food could be found, nonluring hunts were almost equally grim and torturous. Sometimes there was literally no food available to them at all except pine needles and bark. In vain, the Fund and six other conservation groups demanded that the Park Service at least stop their practice of grooming the roads so that the public could enjoy snowmobiling. As it was, one almost totally starved buffalo might be pursued by sometimes dozens of snowmobiles, and many of the drivers seemed, if not to enjoy the buffalo's plight, to be extraordinarily impervious to its now double misery.

By the spring of 1997 more than 2,500 buffalo—

more than two thirds of the total Yellowstone herd—had been killed. And yet the Park Service, the State of Montana, the Department of Livestock, and the Department of Agriculture not only refused to stop the killing, they went right on with it, although this time their cruelty reached new heights. The buffalo were not just shot right away anymore. Instead, they were lured from the Park with bait, stuffed into trucks, and carted off to a slaughterhouse. Here, under a cloak of secrecy, they died in terror. They were shot in the head and, while still kicking, they were hung upside down and their throats were slit.

It was all too much for anyone with a sense of decency, let alone mercy, to bear. And it was all happening to literally the last wild herd of North American buffalo in the whole world. And, again, all because of a disease called brucellosis, despite the fact that there had never been, as we have seen, one documented case of brucellosis being transferred from wild buffalo to cattle.

The Fund for Animals took out a nationwide advertisement expressing our outrage at the butchery, and all over the country there were all kinds of protests. Not the least of these was a meeting in Helena, on the steps of the Capitol, of representatives of forty Indian tribes to mark three special days of prayers for the buffalo. Also in Montana, near Yellowstone, when Governor Racicot, together with two U.S. Senators and the U.S. Department of Agriculture Secretary Dan Glickman, called a press conference at a cafeteria filled with five hundred people, an animal volunteer raced into the room with a five-gallon bucket containing rotted buffalo innards. As she went toward the Governor, one of the Senators' aides tried to grab her arm but all he succeeded in doing was having the whole foul-smelling contents of the bucket go all over the table and splash not only the Governor but also both Senators and Mr. Glickman as well. The Fund for Animals does not usually take part in such activism, and we did not in this case. We did not, however, feel it was unjustified.

The number of letter writers was extraordinary. To those of us who were trying to stop the buffalo killing perhaps the most moving one was from Ms. Gail Cole of West Yellowstone. She wrote us in March 1997:

We had, for a three-week period, a baby buffalo out here staying out on Highway 191 just past the Teepee Creek area. Some kindly people had fed him alfalfa and he just stayed there. He even slept there right on the edge of the highway—on the alfalfa, in fact. He was alone and he was lonely and he didn't know what to do or where to go. We have no speed limit here and I couldn't believe a speeding driver or semi hadn't claimed him for roadkill. He was still within the Park, however, and he could not officially be shot. I called the Park Service and asked them if they knew he was there. They said they were "aware" of him but because he "wasn't endangering anyone" they would leave him alone. They also said he "wouldn't survive anyway."

People had named him Norman. Like some of these people I kept monitoring his daily routine. Several times I saw fast-moving cars and trucks come within inches of him. Finally—it was on a Saturday, I remember—Norman made his move. He got up and started walking back, outside the Park. He was clearly headed for his herd in the Horse Butte area—unaware that all of them had been shot there—bulls, cows, and even calves. After three hours of walking, however, Norman was hit by a speeding pickup and killed. The driver of the pickup, I remember, was very concerned—not about Norman, but because of his pickup. It was, you see, damaged.

Thinking back about Norman, I kept wondering what would have happened if the people who fed him had been caught doing so. They could have been arrested and fined and jailed. They could have been arrested and fined and jailed if they took him to their own land. They could even have been arrested and fined and jailed if they took him to a veterinarian. The only thing they could have done, while Norman was anywhere outside the Park, was to shoot him. Indeed, anyone in the entire sovereign State of Montana would have been faced with similar problems if he or she wanted to do anything to assist the gallant effort to survive by a member of the last free-roaming buffalo herd in North America—a lost, lonely, little stray orphan baby.

Black Beauty Ranch's own buffalo herd consists of one one-eyed bull and three females, as well as a baby female who was born at the height of the buffalo killings. The one-eyed bull had also been a lone baby. He came to us from a volunteer, Laura Yanne, who found him with some cattle in a feedlot in Pennsylvania and had taken him to the Fund. She never knew what had happened to his eye, but in any case we named him One-Eyed Jack. The females were formerly of the Dallas Zoo, who sent them to us when the zoo's "motif," as they expressed it, became African wildlife, and so they phased out their buffalo. Since the baby was born shortly after the Yellowstone massacres, we decided to name her Hope in the then seemingly distant expectation that there would at least be one buffalo with some of same.

When, however, I told Chris that I wanted him to get me a picture of Hope and her mother he informed me that the picture would have to include One-Eyed Jack. I was glad about this, but I was also curious. "One-Eyed Jack," he explained, "would not let anyone on two legs get anywhere near Hope." I told him I could well understand why, but, in any case, Chris and his photographer friend did manage the job—at least at a respectful distance.

When he first came to us, Chris knew virtually nothing about looking after elephants. Nonetheless, in a remarkably short time, he learned an incredible amount, including designing and building what has become regarded as the model, state-of-the-art elephant barn—one in which the elephants, or the people looking after them, are not enclosed in a cement box which neither the elephants nor the people can get out of in an emergency—

but in contrast, one which is built surrounded by poles that have space enough for people to get through. It is only one of the many improvements Chris has developed as he learned to handle elephants, and all of them have been done for the good of elephants as well as people. Today Chris says simply, "I love living with elephants. I don't think I could live without them." He pauses. "Or," he goes on, "without chimps or horses or burros ... or dogs or cats ..." until I stop him.

One thing is certain. The elephants of Black Beauty Ranch are, to all of us, among its most treasured animal inhabitants. Our first elephant rescue was a female named Conga. She came from a place called Jungle Larry's, in Florida, where, along with two other elephants, she was bought at an auction. The other two were sold to a Venezuelan circus and while on a tour with this circus, the truck in which they were traveling fell over a cliff and both were killed.

Conga, an African, is a very large elephant—over ten and a half feet tall and weighing over ten thousand pounds when we first got her—and, though in her twenties, still seems to be growing every year. Before she came to us she apparently had the job at Jungle Larry's of sitting on a park-side bench and twirling an umbrella with her trunk, an act that for an elephant of her size must have seemed utterly ridiculous. When we got her she was none the worse for the umbrella twirling, but there must have been a shortage of shade where she twirled because by the time she came to us her entire skin was sunburned practically down to her feet. As a result, she required not only regular shade but also regular and sometimes even massive skin care. Despite her size and skin problems, however, she has always been a curious combination of both timidity and rambunctiousness and at the same time she is also a combination of skittishness and friendliness. She spooks extremely easily, however, and in one of these spooks is fully capable of hurting herself and anyone else who gets in the way.

Only twice, despite her occasional rambunctiousness, has Conga ever bumped Chris. Once was when she had broken a tusk and Chris was doctoring it. It evidently hurt her and she pushed him, although not hard enough to knock him over. On another occasion, before his state-of-the-art barn was built, she bumped him in what was then her shed and would have pinned him against the wall if he had not moved quickly. Despite the infrequency of these incidents, Chris says, "I still remember. When she moves, I move."

One day when I had been away from the Ranch for some time Conga was so furious I had not come to see her right away that she picked up a fully aired tire from among her playthings and hurled it at me from at least twenty yards away. Yet a moment or two later she was almost cuddling in her affection. "Every morning," Chris says, "we try to evaluate her mood and see whether she is fussing about something—like a storm the previous night or a low-flying airplane—or whether she's positively purring. Even on the day I was fixing her tusk, and it hurt, as I said, she knew I was doing it for her and she never took advantage of me beyond that almost reflexive push. It's the same thing when I give her a bath. At some zoos they make elephants lie down when they have a bath. I don't. At our Ranch we never try to be all-controlling with our animals. I want them to be standing up and close to me when I give them a bath. Lying down, they don't participate in the bath. Standing up, they do."

One day I asked Chris what I should do when Conga sees me from the far end of the pasture and comes running toward me. You know, with this damned arthritis, I told him, I do not move very quickly. What would I do, I asked him, if Conga did not stop? Chris looked at me as if I had asked not only a very stupid question but also one at which he obviously took some offense, as he is inclined to do at any criticism of Conga. "She'll stop," he said crossly. "She knows you can't move fast. She's not stupid—she's an elephant."

The second rescued elephant we had came to us from New England—from a zoo in Pawtucket, Rhode Island named Slater Park. This elephant had the unsuitable name of Fanny—it was unsuitable because she was, even for an elephant, grotesquely overweight. Nor was obesity Fanny's only trouble. The entire Slater Park Zoo was to blame for many more troubles. It was part and parcel of so many zoos built by so many cities and towns in early days that animals existed in solitary confinement, as if they were in jail. This zoo, in fact, had a long history of misery and trouble for the animals. One of the animals, a camel, had killed a keeper who teased it. Afterwards the police, typically, shot the camel. In the same way a polar bear who kept getting loose apparently did so once too often, at least for the police, because they shot him too.

Fanny herself, an Asian elephant, came from the

famed Ringling Brothers and Barnum and Bailey Circus, who had apparently been anxious to get rid of her because they thought she was what circus people call a "runner"—meaning an animal who, on occasion, runs away and whom circus people do not like, because such animals are extremely inconvenient when the circus is paraded down Main Street prior to opening. The Slater Park Zoo had had another elephant—a baby—but the baby was there for a short time only and was being kept, supposedly, for another customer. Aside from the brief period with the baby, Fanny was left without a single other animal companion for thirty-five years. For all but a few months in summer of that time, when she had a small outdoor enclosure, she was confined in a windowless cement barn with a chain around her ankle. When Chris first saw her, after we had learned the zoo was about to be closed and I had asked him to look at her because we might agree to take her, he called me. Well, I asked him, how is she? "She looks," he said, "like an abandoned truck."

Fanny was relatively small in height for an Asian, yet she weighed 9,700 pounds—all too much of it the result of fast foods, doughnuts, marshmallows, popcorn, candy, and other snacks brought in brown bags by people who came to look at her. In the entire zoo there were only three keepers, and the one of these who seemed to be in charge of looking after Fanny seemed to Chris to do most of his "looking after" with a bull hook. Fanny's feet were in truly awful shape, partly because she had had so little opportunity to move around and partly because they were so neglected.

When word that the zoo was closing spread, an extraordinary number of people seemed to want Fanny, even though the Black Beauty Ranch had been one of the first suggested. A particular effort was made by Marine World—Africa USA, who wanted Fanny for, of all things, rides. At forty-seven years old she seemed to us far too old for this and, fortunately for us, Aaron Wishnevski, a prominent Pawtuckian businessman who seemed to be the head of the Pawtucket Committee to Decide the Fate of Fanny, agreed with us—this despite the fact he was given the full red carpet treatment when he went to visit Marine World. Chris, on the other hand, gave Wishnevski very low-key treatment when he came to Black Beauty because, as usual, Chris was busy. When Mr. Wishnevski asked him why he wanted the problem of an elephant like Fanny, Chris told Mr. Wishnevski that is what Black Beauty was for—problem animals—and he liked the challenge of trying to fix Fanny up and making her retirement a happy one. The answer pleased Mr. Wishnevski very much and once he was back from Black Beauty the Pawtucket Committee voted, with just one exception, for Black Beauty. The one exception was the zoo curator.

The week before Fanny's trip to Black Beauty the zoo held a two-day "Fanny Fest," as they called it. At this it seemed everybody in Pawtucket arrived—and of course with all the typical goodies. At Fanny's feet were literally piled not only the usual fast foods, doughnuts, marshmallows, popcorn, candy, and other snacks but also even more ridiculous bon voyage presents. Chris knew, however, that the real trouble would come the day after "Fanny Fest"—on Fanny's getaway day. For this Chris carefully spread the word that moving time would be seven o'clock in the morning. By six o'clock he knew there would be more people than even the number that had come for "Fanny Fest." There was even a rumor that some young people had changed their position on letting Fanny go and were planning to chain themselves to her cage. All this would be happening, surely, by six o'clock. And indeed all this did happen. There was, however, one important change in the proceedings. There was no Fanny. Chris had seen to it that the truck came at midnight the night before.

Arriving at Black Beauty, Chris went to work—not only on Fanny's feet but also on her obesity, colic, and other troubles. Immediately she was put on a diet, an exercise regime, and junk food abstinence. Even her name was changed. There would be no more Fanny. Her new name would be one that we all wanted to have a nice Indian sound to it, fit for an Asian. So, ignoring *Gone with the Wind,* we chose Tara.

At first there was, for Tara, no Conga. Chris had seen to it that Conga was, at first, in a far-off pasture and then afterwards went about bringing them together as carefully and slowly as possible. From the moment Tara first saw Conga she was terrified. All she wanted to do was run off and away and as far from Conga as possible. This went on for four weeks. For half the day Chris would keep Conga in the barn and let Tara out. Then for the other half it would be Conga out and Tara in. And then, finally, came the day when both were let out.

But this time Chris, holding Conga gently by the ear, started walking ever so slowly toward Tara. As Conga approached, Tara would immediately walk away. Equally immediately Chris would follow. As I watched, it was as if they were having their own little parade. Finally Tara, who still moved slowly, seemed to tire of the parade and stopped in her tracks. This time when Conga came right up Tara did not move away from her but instead just stood there and then, of all things, first touched, then sniffed and finally seemed to hug her. Tara actually trumpeted with excitement—the first trumpeting we had heard her make since her arrival at the Ranch. Then, copying Conga, who held her head up, Tara too held her head up, higher than it had ever been so far at the Ranch. And this time, side by side, they went for a swim in what was now not Conga's pool, but their pool. For the first time in thirty-five years Tara had a friend.

The third rescued elephant came to us after Tara, but also from a circus—in fact from a circus which had eight elephants and in which she was not only doing all the acts but doing them better than all the other elephants. But she too had troubles. She had terrible trouble both with her right front leg and also her right rear leg. And, according to the circus people who brought her to us, although people did not notice her disabilities when she did her act in the circus, they could not help noticing it when she dragged her legs—which she could not help—during the parade. On top of all this, her head was not right—it was misshapen.

There was good reason for all this. Babe had come originally from South Africa and was literally standing beside her mother when there was what they call a "culling"—meaning, of course, a killing—and her mother was actually killed right beside her. She came to America, and the circus, by ship, in a crate, and it was in the crate that, we were told, she not only banged up two of her legs so badly but also, as evidenced by her head, had tried to commit suicide. This was easy to believe because, by the time she came to us, her head was still so misshapen it was hard to believe anyone would have tried to do anything except help her get better, let alone put her in a circus.

Despite her appearance and her disabilities and her obvious depression, under Chris's almost round-the-clock attention Babe made remarkable improvement and even before she did, the way she tried to get over what had obviously been for so long all her troubles was so winning that she became almost immediately the most popular animal we have ever had at Black Beauty. Now ten years old as I write this, she is so full of spirit and mischief and affection for all creatures, animal and human, that even some grumpy grown-ups who take their children on a tour of the Ranch but who do not seem to care for animals very much themselves cannot resist Babe and always end up wanting to know all about her.

Babe did not, of course, get to this special position right away. Everybody at the Ranch was, for some time, occupied with trying to do things for her. She had her own special stall built in the elephant barn, almost invariably wore a blanket even when it was warm, and had a huge helping of wood shavings on her floor, to protect her feet. Way too skinny at first, and seeming to have cuts on various places all over her, she also had obviously itchy, dry skin which had to be oiled every day, and sometimes several times a day. She was given a bath at least once a day, and during this hydrotherapy treatment had warm water running over all parts of her body, but particularly her legs. We all played with her constantly, and she with us. Indeed, we sometimes felt she had never really played before she came to us and, we soon learned, we were correct. Elephants in circuses are not allowed to play. The trouble is, you see, they get their costumes dirty.

At first Chris did his best to keep Babe away from Conga and Tara. But from far across the elephant pastures Conga spotted her and immediately became extremely interested. And suddenly, one day, to Babe's own little stall, Conga brought the newcomer some of her own hay and grain—although she already had plenty of both. Among her other troubles Babe had only one tusk. This, however, was a very long one and when she swung it quickly it was extremely dangerous for all concerned. Finally the day came when Chris decided that he would have to trim the tusk. As he started to do so, however, Conga set up such a ruckus that she had to be forcibly removed from the proceedings. In the end, though, Babe's tusk grew back—as it should—and not so long, and even Conga approved.

Tara was another matter. Tara thought nothing of Babe. Whether this was because she had had a bad experience with the one other elephant in her life—the baby who had been put with her for a time, and then taken away—we do not know. It could even have been that she was jealous of an animal that might have seemed to her to take Conga away from her. In any case, for a long time, Tara was so upset by Babe she would not even eat when Babe was nearby. If Babe stuck her head out of the elephant barn window and Tara, outside, could see her, Tara would immediately run to the farthest end of the pasture. Then, as quickly as Tara's dissatisfaction with the new tenant had begun, it suddenly stopped. And one day we looked out at the pasture, and there were not only Babe and Conga, but Babe and Conga and Tara—all three with their trunks entwined with each other.

Almost from the beginning Babe, of all our animals, was the only one given the run of the Ranch. Chris started it by taking her to various places on the Ranch he felt she would find interesting. And, in true elephant fashion, they were the places where Babe would first go and check out when she was on her own. I turned out to be the real worrier about this—about Babe when she was out, on the theory that nobody at the Ranch would be looking out to see that she had not gone somewhere, or perhaps had even run down the road. But I really didn't need to have been that much of a worrywart. Whoever first let her out on a given day always kept a weather eye out for her and, in time, Babe, more so and more quickly than any animal at the Ranch, mastered all the Ranch locks and gates. She even went so far as to have night slumber parties with whatever animals struck her fancy of an evening—from burros to buffalo. On one occasion she decided that the burros would rather be with the horses than just with the buffalo, so she opened their gate, the first time to let the buffalo go to the horses, and then another time to let the horses go to the burros. Many times, watching Babe, I have the feeling she not only has the run of the Ranch, she also runs the Ranch.

All animals need to be talked to, Chris feels, but elephants need it especially. The best way, he says, to get them to do something is first explain to them what it is you want them to do or where you want them to go and, then, after you have talked to them about it, start moving them or getting them to do what you want by nudging them or pulling them gently on the ear. Actually, Chris's talking when he is working with the elephants is almost constant and when he says the elephants understand almost everything he says, he means it. "It's not really," he said, "the way the chimps understand, but it's close. Elephants are very social wild animals, perhaps the most social of them all, and they want to understand, even if they don't completely."

ALL IN ALL, CONSIDERING THE WONDERFUL QUALITIES of elephants and their remarkable closeness to and affection for humans, it is truly incredible that so many of them are treated so cruelly by humans. With all three of our elephants we had three different examples of this, all from their own particular areas of cruelty—Conga, from a roadside zoo; Tara, from a city zoo; and Babe, from a circus. Ever since I was little I had always loved elephants—long before I knew anything about cruelty to them. And even when I was somewhat older, I still knew very little. But the time would come, of course, when I would know more than I ever wanted to know, and this came, most ferociously, in 1988 in the matter of an elephant named Dunda at the San Diego Zoo. It is something that I doubt will ever be forgotten by anyone who cares about animals.

There were two reasons why this was so. One was that it was such an unforgivably outrageous cruelty visited upon an animal that it would have reverberations about their treatment for years to come. The second was that it happened in a zoo which, at least up to the time of that incident, was not only highly respected but the first or second in terms of respect in the entire country. Because of one incredible cruelty, however, it will be a long time before the San Diego Zoo will ever attain that position again.

Dunda was a young, shy, timid, lonely female. She was born in Zimbabwe, then she became an inhabitant of the main San Diego Zoo and she was apparently reasonably happy there. But she was suddenly removed from there and taken off some distance away to what was the zoo's so-called Wild Animal Park. Here she was not only parted from her friends but also put in the zoo's so-called breeding program—something they were very proud of since it not only produced baby elephants to be shown off to the public but also produced them at no cost. In any case Dunda, disoriented and now even more lonely, did not do well in the Wild Animal Park and, one day, even committed a zoo cardinal sin—she did not actually strike one of her keepers, but she swung her head around and seemed as if she was going to lunge at him.

The next day Alan Roocroft, the director of elephants at both the zoo and the Wild Animal Park, ordered what he called a "disciplinary session." What happened at this session—most of which did not become public for many days—was that, the day after her alleged lunging, Dunda was stretched out with a block and tackle and, while Mr. Roocroft yelled voice commands, other keepers—two on each side of her head—beat her with axe handles, first the first two and then, in relays, two more, all of whom beat not only her body, but particularly her head. This went on during one morning and then again for another session in the afternoon, and another the day after that, again in the morning and in the afternoon. At the end she let out first a dreadful moan and then finally an even more dreadful gasp. "I thought," said one keeper who participated in the beating, "that she was about to die, and I'm sure if we had struck her one more time she would not have come back."

After the beating had finally stopped, Dunda was pulled up—again by block and tackle—and was given an apple.

Just how the zoo was able to cover up the whole awful story would seem almost as incredible as the story itself, unless one was familiar with both the zoo and its highly organized publicity network. Even Dunda herself, her head a mask of welts and bruises, was hidden from any outsiders' sight. Finally, however, primarily because of two of the zoo's elephant keepers, Steve Friedlund and Lisa Landres, the story did break and when it broke, it broke in all particulars, complete with pictures from insiders.

From here on the hero of the story was a man named Dan McCorquodale, a remarkable California State Senator and chairman of the state's Natural Resources and Wildlife Committee. Despite opposition from not only the San Diego Zoo, which tried in vain to give the Senator one of its whitewash publicity tours, but also virtually every other zoo in the state, McCorquodale demanded, and soon chaired, in Escondido, a hearing on what had now become "The Dunda Case."

The zoo brought out witness after witness, from their board members to their curators and from their veterinarians to their publicity icon, Joan Embry, all of whom had a vested interest in the cover-up and all of whom, it seems, made three claims—that the beating had not been that severe, that severe discipline was necessary in the handling of elephants, and that the Dunda case was an isolated incident. None of these claims was substantiated and, frankly, none of them was true. As for the beating of Dunda being an isolated incident, it was all too apparent that such beatings, at that zoo and almost all others, were routine, if for no other reason than that they made it easier to get the elephants to perform tricks. One of Mr. Roocroft's favorite tricks, for example, was to conclude a Wild Animal Park tour by having the visitors see him lie down on the ground, on his back, and having an elephant stand over him with one foot just over his face.

When the hearing assembled in Escondido, it was soon clear that the zoo was not only backing Mr. Roocroft and his beaters but was also pulling out all the stops to do so. Not only did they take all the chairs in front of the committee panel but they also tried every way they could to keep people on the Dunda side from testifying at all. I personally felt more pressure than I had ever felt at an animal hearing, including being called on the telephone by Betty Jo Williams, chairman of the zoo's board, and virtually begged not to appear. As if this were not enough, just before I was introduced, Mr. Roocroft himself came over to me and told me that he had been "authorized" to offer me something. He then explained what this offer would be: He would, at the zoo's expense, take his best bull elephant, as he put it, to "your Black Beauty Ranch" and, "no matter how long it took," would breed him to Conga, and that our Ranch could have the baby. I was so flabbergasted by the offer that I had no idea how to answer it until McCorquodale called the meeting to order. This somehow galvanized me to action and in a voice that I hoped would be heard by the Senator and at least some of his other committee members, I told Mr. Roocroft that the Fund for Animals had never been in favor of breeding animals, but that now, just knowing him, I was not at all sure we were in favor of breeding people.

One particularly memorable testimony about the Dunda case came after the committee hearings by Ray Ryan, a longtime San Diego Zoo elephant keeper, who was asked by a local radio show if he would be willing to talk about what had been going on at the zoo and the Wildlife Park. Mr. Ryan agreed but when he arrived for the show the host of the radio show said that zoo officials had informed him that if the station ran his talk show all the zoo's sponsorship, which was apparently large, would be immediately pulled. Just the same, the host decided to go ahead with the interview anyway. After the show, however, zoo officials invited the host to meet Mr. Roocroft and the elephants up close to see for himself how well the elephants were treated. When Mr. Ryan asked to accompany the host, to balance things out, his request was of course refused. The host was then treated to a tour of the Park by Mr. Roocroft himself and by the time he left he was, in Mr. Ryan's words, "a little kid in the palms of their hands." Even then, Mr. Ryan noticed the station replaced the host anyway, and then and only then did the zoo resume their advertising.

It was too bad Mr. Ryan had not had the chance to tell the committee about the way Bisi, a shy little elephant who was also from Zimbabwe and was a close friend of Dunda, was taught by the San Diego Zoo how to stand on her back legs. Mr. Ryan's story was as follows:

> *To get her to stand on her hind legs, like you see in all the elephant shows, there was a harness that was put around her front legs which was attached to a block and tackle. The block and tackle went over the top of a small storage barn. Bisi was backed up to the wall of the barn, and as several keepers pulled on the*

rope Bisi's front end was slowly lifted off the ground. If and when she tried to pull herself down, she was beaten on either side by a keeper with an axe handle to force her to stay up on her hind legs. This was how all elephants were trained to stand on their hind legs. Lou said that Bisi would be screaming at the top of her lungs and her legs would be shaking violently every time this was done. She never seemed to get it right, and during every show when it was time for her to perform, she would defecate all over herself. After the show when everyone had left, she would be brought out again and made to go through the routine over and over until she got it right.

It is interesting to note that Bisi and Dunda became lasting friends. It is also interesting, in view of Dunda's being given an apple after her beating at the Wildlife Park, that some time after Dunda's beating a new woman keeper came out among the zoo elephants with an apple in her pocket to give to the elephants. Unfortunately, the apple fell out of her pocket and the elephants got so excited while trying to get it that the zookeeper was trampled to death. That, too, was covered up as much as possible. In the end, the zoo got rid of Dunda by sending her to the Oakland Zoo and then proceeded to see that she and her friend Bisi were parted for good by sending Bisi to the Waco, Texas, zoo, where she died.

In any case, the credit for making the Dunda case a lasting issue not only in California, but also nationally, belonged to Senator McCorquodale. Once he had finished his San Diego Zoo hearing at Escondido, he moved on to hearings at many other zoos in the state.

One of these took place at the Los Angeles Zoo and here this zoo's director, Warren Thomas, had even outdone San Diego in trying to ensure that a huge contingent of people would be on his side. He had a large number of them in the zoo's large auditorium and orchestrated them not only to keep demanding to speak on the zoo's side but also, before anybody on our side spoke, to jeer and hoot at us so much that we could not be heard. Only with great difficulty and a good deal of luck was I finally able to get an opportunity to speak, at which point I felt I had time to get one short question through to Thomas before the jeering and the hooting would make my effort meaningless. Sir, I asked Thomas during this merciful pause, do you just get a kick out of beating elephants or do you like to beat other animals too—like, say, rhinoceroses?

McCorquodale's bill was finally passed in the California legislature:

It shall be a misdemeanor for any owner or manager of an elephant to engage in abusive behavior toward the elephant, which behavior shall include:

(a) Deprivation of food, water, or rest.

(b) Use of electricity.

(c) Physical punishment resulting in damage, scarring, or breakage of skin.

(d) Insertion of any instrument into any bodily orifice.

(e) Use of martingales.

(f) Use of block and tackle.

At the last minute, two parts of the bill, which had at first been included, were removed. One was that chains should not be used as the principal method of restraining or confining an elephant. The second was that members of the public would not be allowed to ride the elephants. These were removed by the incredible persistence of the lawyers and lobbyists from Ringling Brothers.

These same lawyers and lobbyists fought virtually every new effort to strengthen the bill. Their method included, at the beginning, the statement that "Ringling Brothers has been handling animals for approximately 119 years and consequently is regarded as being one of the world's leaders in both experience and knowledge regarding the needs and successful husbandry of animals." After this mouthful, the next paragraph, apparently equally meaningfully for the legislators to hear, stated that "Ringling Brothers spends approximately two and one half months in California, providing spectacular family entertainment to hundreds of thousands of families in California."

Next the lawyers and lobbyists would go after any specifics in any proposed bills. For example, on the subject of chaining, they stated, "There is absolutely no scientific evidence or research which shows that the chaining of an elephant is unsafe or detrimental to the elephant." Following chaining, Ringling Brothers eschewed the idea the animal people had suggested that "every elephant shall have daily access to a sizable mudbath area or pool and dirt, for dusting." "The idea," they responded, "that an elephant have daily access to a sizable mudbath area or pool and dirt, for dusting, has no scientific basis nor is it sound animal husbandry."

Ringling's next objection was to the idea that "no fewer than two elephants shall be housed together and larger groups shall be encouraged." For this, once again, the lawyers and lobbyists demanded proof and stated, "There is absolutely no proof that an elephant is harmed if it is not housed with another elephant." Such wording, it should be noted, could hardly pass muster with any animal person if for no other reason than such a person would not permit the use of the word *it* for an elephant but would insist instead upon either *he* or *she*.

IT SHOULD BE REMEMBERED THAT AT THE TIME SUCH problems were being debated in the legislature, Florence Lambert, of the Elephant Alliance in La Jolla, California, was busy preparing what she called, simply, "Death List." It read as follows:

HUMAN DEATHS CAUSED BY CAPTIVE ELEPHANTS

1990–1996

1990	Ft. Lauderdale, Fla.	Trainer killed by elephant owned by Hanneford Family Circus.
	Loxahatchee, Fla.	Handler at Lion Country Safari killed by elephant.
	Japan	Keeper killed by elephant at Cunna Safari Park.
1991	San Diego, Calif.	Keeper killed by elephant at the San Diego Wild Animal Park.
	Tokyo, Japan	Keeper killed by elephant at the Kiryu Zoo.
	United Kingdom	Keeper killed by elephant at Twycross Zoo.
	Windsor, Canada	Circus hand killed by elephant.
	Oakland, Calif.	Worker killed by elephant at Oakland Zoo.
1992	San Antonio, Tex.	Keeper killed by elephant at San Antonio Zoo.
	El Salvador	Keeper killed by elephant at National Zoo.
	Moscow, Russia	Keeper killed at the Moscow Zoo.
1993	Fishkill, N.Y.	Man killed by a Clyde Beatty–Cole Bros. Circus elephant.

1993	Tampa, Fla.	Keeper killed by elephant at Lowry Park Zoo.
	Willston, Fla.	Head elephant trainer killed at Ringling Bros. "breeding compound."
1994	Honolulu, Hawaii	Keeper killed at Circus International by elephant owned by Hawthorne Corp.
	Denmark	One person killed by circus elephant.
1995	Belgium	Keeper killed by two elephants at wildlife park.
	Thailand	Two mahouts killed by circus elephant.
1996	Thailand	Man teasing elephant killed by elephant.
	Santiago, Chile	10-year-old boy killed by circus elephant.
	Rome, Italy	Trainer killed by Tongi Circus elephant.

CIRCUS ELEPHANT DEATHS

1994–1997

Name	*Age*	*Sex*	*Date*	*Cause of Death*
Assam	24	M	June '94	Died under "sedation."
Siam	?	F	July '94	"Euthanized"
Tyke	21	F	Aug. '94	Shot
Amy	?	F	Sept. '94	Unknown
Kay	50	F	Oct. '94	Kidney ailment
Mona	26	F	Oct. '94	Metabolic disease
Dumbo	43	F	Dec. '94	Tuberculosis
Bombay	40+	F	1994	Unknown
Sahib	20	M	Jan. '95	Killed because elephant was "unmanage-able."
Lois	24	F	Jan. '95	"Foot infec-tions that had spread through her body"
Mike	?	M	Mar. '95	Burned to death
Rhonda	?	F	Mar. '95	Burned to death
Jockey	20	M	Mar. '95	Shot

Mary	50+	F	Apr. '95	"Collapsed and died."
Stoney	21	M	Aug. '95	Complications from training injury
Bandula	30	F	June '96	"Euthanized." Suffered 23 years with arthritis.
Joyce	47	F	Aug. '96	Tuberculosis
Hattie	27	F	Aug. '96	Tuberculosis
Tunga	32	M	Sept. '96	Unknown
Ola	?	F	Mar. '97	Post foot surgery

As in the Dunda situation, there was one particular case that stood out above all the others. This was the case of the elephant Tyke who, working with Circus International in Honolulu on August 23, 1994, first kicked her groom, William Beckman, then crushed her trainer, Allen Campbell, to death, and finally broke free before she was caught, cornered, and shot at by what seemed like an army of police. Even then she was killed not by the bullets but by lethal injection.

The late Carl Viti, the photographer for the *Honolulu*

Advertiser who took the most memorable pictures of Tyke, wrote a letter to Mrs. Lambert expressing his feelings:

Not one day goes by that I don't think of that poor elephant. The cops counted eighty rounds fired by their people. I saw at least twenty rounds hit Tyke. She was probably hit at least fifty times. At that time she could only lift herself on her front feet and could only swing her trunk pathetically, trying to ward off the bullets. It was the sickest thing I have ever seen, and I've been shooting news for over twenty years. I just feel empty inside, like something is missing.

Tyke was owned by John Cuneo, of the Hawthorne Corporation. She was, it turned out, what circus and zoo people called, as we have said, a "runner." Before the Honolulu horror she had run from the circus no fewer than three times—all three in the very year before Honolulu. In April 1993 at the Altoona, Pennsylvania, circus she had even ripped away part of a wall to get out and away. In June, at the Harrisburg, Pennsylvania, circus, she had bolted out and partly through the circus tent, and in July, in Minot, at the North Dakota State Fair, she broke two of a show worker's ribs, then bolted from her trainers and led them and Ward County Sheriff deputies on a twenty-five-minute chase through a street that runs parallel to the midway.

One would imagine after all the evidence of cruelty and death to both elephants and people that one would think twice about exhibiting elephants in zoos and circuses, but less than a year after the Tyke tragedy, something called "A Ringling Brothers and Barnum and Bailey Circus Anniversary Commemoration" was held on, of all places, the Capitol grounds. This was the event not only of Barnum and Bailey publicists but also of Newt Gingrich, who apparently saw in the promotion of elephants on the Capitol grounds some sort of firmer footing for Republicans in the future as long as there were only a few Democrats or perhaps donkeys present.

In any case, whoever's idea it was, in the entire United States Senate there was apparently only one voice of reason that was raised to protest this ill-advised event. The voice belonged to Senator Robert Smith, Republican of New Hampshire, and his amendment on the whole idea of having elephants on the Capitol grounds was a simple one: "No elephant," it said, "shall be allowed on the Capitol grounds for the purpose of this event." Only a handful of people turned out to hear Senator Smith's words, but those who did found in them a moving eloquence rarely heard in Congress on behalf of any animal cause. The following words are from the Senator's closing remarks:

I want to stress a few points in closing here. Ringling Brothers maintains their training practices are not cruel and they are not abusive. I think they mean that. They may think that.

But let me say, when the elephants go berserk, the first person they go after is the trainer. That ought to say something. When I met with Mr. Ireland he said that while Ringling Brothers does in fact use whips, whips are used because of the cracking sound, which is an audible cue for the elephant.

I am not an elephant trainer. I do not know what the function of a whip is, or how it works. I suppose if someone was whipping a cracking whip behind me, I would do whatever they said, too.

I have concerns about a number of other practices that are regularly employed in the training of elephants. I am not going to get into whether Ringling Brothers employs these or not. I do not know. We may never know, because no outside organizations are allowed to monitor or film their trainers.

I was offered the opportunity to go down to a Ringling training area where they train elephants by Ringling. They said, "You can come in and watch us train." I found that somewhat humorous. If they had any methods I would object to, I do not think they would use them while I was there. Maybe they would, but I doubt it.

These tapes, I have them. I would be more than happy to provide them for my colleagues to look at them anytime they want to look at them. The hooks that are used, the methods of training the animals "down," the cramped quarters to house the animals, and the methods used in breaking wild elephants.

Let me just say for the record on wild elephants. Ringling has assured me that they do not use wild animals, that they breed their own and take young elephants, and I have no reason to deny that.

In the past and, in fact, in some other circuses, baby elephants are captured in the wild, taken from their mother, and beaten for days at a time while they are screaming. It is on the tape. Members ought to watch it. Beaten, for days and days, in shifts, by these people in the jungles of Thailand and Laos. Whenever the elephants are captured, beaten consistently until their spirit is broken, and until they have nothing left to offer resistance to. Screaming and crying. Ought to watch the tape.

The issue, really, is this. Should an animal this big, this wild, be used for entertainment? I do not think so. I do not see the need for it. There is no need for it. We do not see what happens when the circus is not around.

We have seen the animal out there with its trunk around another elephant's tail and gets up and does a trick. That looks cute, and I have seen it. Frankly, before I knew more about this I thought it was great. How do they transport an elephant from Florida to California? You cannot put them in an airplane so they put them in some kind of a truck. Ever look at the width of a highway? There is only so much size of a truck that can be used.

So there are cramped quarters. Now, when you have them on location for a circus—let us say, down here at the Armory—how do you restrain these animals? How do you restrain an animal that weighs several tons? Let me tell you how you restrain them. You chain them. You chain them up.

You can say we feed them hay, we feed them grain, we take good care of them—these are wild animals, so that is why things like this happen. That is why elephants go berserk, because they are not meant to do these kinds of things, and it is cruel. It is cruel.

We have an opportunity here today in the Senate to make a very small statement. We are not going to stop this, but we could say, as U.S. Senators and U.S. Congressmen, that we do not want to risk having an incident like this happen on the Capitol grounds, No. 1 and, No. 2, we think that, even though it is not intended to be cruel, the result is that it is cruel, in the way we treat these animals. They ought to be left alone, in zoos and in parks and wherever we can, and not use—or abuse—them in the way that it is being done in these circuses.

But they do not have anybody. There is nobody who can come out here on the floor. An elephant cannot come out here on the floor. No animal can come out here. It does not have any Congressmen or Senators to represent it. So if somebody does not speak up, who do they have?

I have made my case. I think I have told the world, the Senate, and hopefully many families and children out there who may want to be coming to the circus—I hope, frankly, you do not. I hope you send a statement that this is wrong and we ought not to do it and we ought to be somewhat considerate, in a very small way, by saying this is wrong.

Senator Smith's amendment was not passed—in fact, it received just one vote in its favor, and that was Senator Smith's. And the circus and the elephants went on as scheduled on the Capitol grounds. But somehow, even with just one vote, Senator Smith had made a point that is almost certain to be remembered long after the Ringling Brothers on Capitol grounds are thankfully forgotten.

THERE ARE MANY OTHER ANIMAL RESCUEES AT BLACK Beauty—indeed, there is a goodly number most people would not know by sight, let alone by what it was. Our aoudad are perhaps the most interesting of these. They are sheep, but they look more like high-performance goats. They can jump as high as ten feet, and they are proud and extremely territorial. Equally interesting and cause for much curiosity are our kinkajous, who people like to refer to as "kinkawho's" because they cannot figure out whether they are monkeys or what they are. Actually, although they climb like monkeys they are not monkeys; they are a member of the wolverine family, although they are vegetarians so they are not ferocious like wolverines. They are also related to the raccoon family and are also called, if you are not totally confused by now, "honey bears." Over the years we have collected three of these, one of whom was found walking in the street in Detroit with a collar around his neck like a dog; although we tried hard to locate his owner, we have never been successful. The animal has also been declawed, something we detest being done to any animal but particularly a wild animal. But somehow, even declawed, he can climb almost as well as his two kinkafriends, who are somewhat strange but we never call them kinky.

Among the animals whom we might have expected not to be especially intriguing to visitors but have turned out to be very much so are our wild pigs. They were part of our rescue, from San Clemente Island, of the wild goats. Tourists find them fascinating because they are so different from domestic pigs. We warn our tourists, however, to look but not touch. An irritated wild pig, several of the staff have already found, has various ways of showing his or her irritation and not the least effective of these is a quick butt of the snout followed by an equally quick tusk butt to your butt. Fortunately we have had few of these butts of a serious nature, but at one time we had so much trouble with poachers shooting at our wild pigs at night that we had to move them to a safer pasture—one patrolled by, of all our animals, our llamas. Our burros are our best patrollers against coyotes—they are especially good guardians of our goats, which is particularly remarkable because burros are not overly fond of goats, which they feel are none too bright or at least not in their league of brightness. Nonetheless, burros like to have a job, and if guarding is the job they will do it for any animal, even ones that are relatively low on the friendliness scale.

Llamas are almost as efficient as burros as guardians and, where the wild pigs are concerned, they will run poachers off the pasture so fast they will know they have never run that fast before. We have three llamas now, and we named all of them properly—Lloyd, Llewellyn, and even L.L.D. We once had four but unfortunately one passed on before we could find an Ll name for him. Curiously, llamas are particularly picky about their animal friends—so much so that not one of ours has really made close friends with another. They will choose as a friend either a horse, a burro, a buffalo, or even an elephant—but not another llama.

Not by any means last in appeal among our animal rescuees are our foxes. By far the majority of these have come and gone, along with our raccoons, when they have recovered from a broken leg, or whatever, and have been released after treatment to our wildlife area, which is now more than a quarter of the whole Ranch. Two arctic foxes, however—one of which is perhaps the most beautiful animal on the whole Ranch, and both of whom were rescued from a fur ranch—have become so close to us that we cannot bear to part with them, even to the well-protected wildlife area. Besides these we have nine Siberian foxes, also rescued from a fur ranch, who have become even closer to us than the arctics—so much so that we are concerned that if we tried to take them to the wildlife area they would be back at the Ranch before we were.

PERHAPS THE MOST UNUSUAL BUILDING ON THE ENTIRE Black Beauty Ranch is a round, six-foot-high edifice which looks for all the world like a huge children's sandbox. Overhung by a large oak tree which gives it generous shade, it is Black Beauty's idea of a very special living monument to one of the most fascinating creatures in the entire world—yet one that is uniquely and solely a North American mammal—the prairie dog. If there is a more charming creature anywhere on God's Earth than the prairie dog, I have yet to have had the pleasure of meeting him or her.

Standing on their penguinlike feet, less than a foot in height and weighing from a pound to perhaps three pounds, prairie dogs are plump little creatures, yellowish-gray in color, and have very short back legs but certainly more active front paws. They have small ears but high-set eyes, with extremely keen sight. Above all, they are almost unbelievably social, playful, and friendly with both humans and other creatures. They love company and literally jump for joy and emit their special high barks just for the fun of being with another of their number. They also use the same quick high bark—emitted twice—to warn of the approach of an unfriendly predator. Even having just three prairie dogs at Black Beauty we have noticed that when they are not in one of their burrows there will be at least one of them doing sentry duty, staring in rigid attention with his or her eyes alternately roving the ground and the sky. In prairie dog "towns," as they are called, at the sight or sound of anything unusual, these sentries will give out warnings—two sharp barks—and a visiting prairie dog will race to his or her own mound and stand upright as if, for all the world, martial law had been declared and the troops had been called out for duty.

Almost everyone who has ever come in contact with a prairie dog town has remarked on what superb engineers these remarkable little creatures are. Dorothy and Lewis Nordyke, veteran prairie dog observers, describe the dogs' burrows as about four inches in diameter, then going straight down for more than twenty feet, thus making what they call a "plunge hole." Other animals, they maintain, are content to race into a sloping burrow but the prairie dogs want to dive deep into the earth when danger is near. A few feet below the surface, the Nordykes point out, is a niche, or "barking place." If the danger alarm is not of an emergency nature, the animal stops here and barks irritably. At the bottom the hole makes a sharp turn, runs horizontally for eight to thirty feet, then slants gradually up to the escape hatch.

In *The Friendly Nuisance,* the Nordykes noted other remarkable facts about the burrows and their mounds:

They were as neat as pins, and sloped spaciously. Some of the prairie dogs seemed to be using their mounds as front porches, nonchalantly sitting there on their haunches and leaning back on their tails. Others were standing bolt upright.

A single animal moved to an outlying burrow and barked. Another emerged from the hole and the two wandered away together. We saw seven dogs gather at one mound, sit on their haunches, and bark as if they were talking. Within ten minutes at least twenty other dogs came to the pow-wow.

Elsewhere in the town there was vivacious activity; bright-eyed animals strolling and bowing, the young rolling and tumbling. When two females met away from their own burrows they bowed and then, standing straight up with their forepaws touching, they put their mouths together, as if kissing. A male and female did the same thing, but when two old males came together they whirled and tried to cover each other with kicked-up dirt.

Our males at Black Beauty—we have two, and one female—do not try to cover each other with kicked-up dirt, but we feel this is perhaps because the female keeps them in line. Jack Scott and Ozzie Sweet, in their book *Little Dogs of the Prairie,* speak of the prairie dogs' gathering in clans. These are, the authors note, far from human war clans:

> *The clans live mainly in peace, with much sunbathing and grooming, often standing or sitting side by side with their forelegs around one another. Clan members are identified by touching noses and sometimes kissing. Tribal in nature, the clan offers complete community cooperation. The members share burrows in time of danger, defend one another from strangers, share sentry duties, and generally live a tranquil life of respect for one another.*

That such friendly, peaceful animals should be persecuted by man at all is sad enough. But not only were they so persecuted, their persecution was almost beyond belief—rivaling that suffered by the buffalo and the wild horse, the wolf and the coyote. It began, as do so many horrors of the West, with the ranchers. Defying the facts, they believed that the prairie dogs were eating too much grass which belonged to their cattle and sheep, and that when they were not doing that they were digging holes into which their horses tripped. Neither argument stands against the facts. The fact that the prairie dogs ate the grass was actually one of the reasons the grass was better for the ranchers. Studies have shown that prairie dogs brought to the surface as much as five tons of subsoil per acre, and their burrows also helped water infiltrate to lower levels and conserved moisture, thus augmenting water tables. As for the ranchers' horses stepping into the prairie dogs' holes, this was a relatively minor matter. A careful horse, even a careful horse without a rider, steps around them.

Lynn Jacobs, in his remarkable book *Waste of the West,* best tells the story of the ranchers' determination to exterminate their tiny, inoffensive enemy. At first, he points out, the ranchers killed prairie dogs with the weapons at hand—guns, poison, traps, dogs, and even disease—the disease being the transportation of sick, plague-infested prairie dogs in trucks, sometimes across hundreds of miles. The use of poisons and diseases, which began in the late 1800s and continued to the next century, was augmented by the stockmen, as they always did, turning to the taxpayers. In short order Federal, state, and local government agents assaulted prairie dog ranges with massive amounts of various poisons. The more they killed, Jacobs states, the stronger grew the stockmen's desire for profits. By the 1920s, urged on by the government agents who also stood to gain, the stockmen were preaching the total extermination of the prairie dog. Jacobs continues the horror:

> *After World War II, the prairie dog "control" program became a lustful, massive campaign of genocide against these peaceful creatures. Compound 1080 was added to the arsenal. Poison was used on all prairie dog towns wherever livestock grazed—in other words, nearly everywhere. Aircraft flew over their vast towns, broadcasting tons of poisoned bait. Soon, all the great colonies were destroyed. Billions of gophers, squirrels, rabbits, mice, seed-eating birds, insects, and microbes died along with the prairie dogs, as did the predators and scavengers that ate their toxic bodies. Some of the poison washed into waterways; some adhered to vegetation and was eaten by livestock. And, interestingly, the lack of rodents caused many larger predators to prey on livestock.*

"Just a hundred and fifty years ago," Jacobs concludes, "five to ten *billion* prairie dogs occupied more than 600,000 square miles—an area over three times the size of California. In other words, there were more prairie dogs in the West at that time than there are now humans on Earth." At the same time, during the campaign against the prairie dog after World War II, more than 800,000 square miles of the West were poisoned.

To this day there are rural communities in various western states that proudly promote prairie dog shoots. One group which boasts members in forty countries is called the Varmint Militia, and their organizers promote their shoots by giving away a copy of the *Exploding Varmints* video to each contestant. For some time the most famous of all prairie dog shoots was held in Nucla, Colorado, which called its shoot "The World Prairie Dog Shoot." I attended one of these and watched the sickening sight of gun nuts lying on their stomachs, with high-powered rifles, with portable shooting tables that pivot 360 degrees and have a resting stool for the rifle, picking off prairie dogs one by one as they popped up from their burrows. The whole scene was a grotesque shooting arcade, all to win a contest to see who could kill the most of the tiny creatures. Along with many other activists we took part in a protest which eventually resulted in Governer Roy Romer himself protesting the shoot and urging them to be stopped. The actual reason for the final ending of the shooting, however, was that Nucla, the town that had become world-famous for its cruelty to prairie dogs, only gave up the sport because there were no more prairie dogs to shoot.

I was reminded of the fact that not long ago E. G. Pope, a U.S. Fish and Wildlife Service agent, stated that he believed all the prairie dog troubles had begun with their name. "If they had been called anything else," Pope stated in typical Fish and Wildlife fashion, "they would have been a fine game animal. They would have been an eating animal, and they would have been controlled instead of exterminated."

Far more to the point, I felt, was a story given me by Michael Markarian, the Fund's Campaign Director—one which contained the comment of Kirk Knox, a reporter and columnist for the *Wyoming Eagle* and the *Wyoming State Tribune*. Mr. Knox stated that he had been reared around and with guns and that once, before he was even teenaged, he had shot a prairie dog. "As I considered this nonsense in Colorado," he wrote, "I thought that if there was but one bullet I had fired that I could call back, that would be it."

L'ENVOI

A GREAT DEAL HAS HAPPENED TO AND ON the Ranch since that day, many years ago, when Peg, the three-legged cat, who is still with us, came up the driveway, still struggling in a leghold trap. The Ranch itself, which started with a nervous buy of eighty-five acres, is now, including our leased land, over a thousand acres strong. And, best of all, where once just a few animals lived there are now close to six hundred, and literally hundreds more who have passed peacefully away with us, but who, before they did, found, as Black Beauty himself had found and the sign over the Ranch says, their home.

Some, it is true, have had just short little lives with

us—so many animals seem to have them. But however much time they have, we want it to be as special as we can make it. Michael Kilian, author, *Chicago Tribune* columnist, and a member of our board of directors, recently visited the Ranch. When he came back, I asked him to tell me what he liked best about it. He paused only a moment. "I have never in my life," he said, "seen so many happy animals."

No comment has meant more to me than that, and I hope any of you who visit the Ranch will feel the same way as Mr. Kilian did. I also hope that you will pardon me for having given you, in this book, such a large dose of animal unhappiness. The plain fact is I did not see any other way to write it, because what I most wanted the book to do was to show what life could be like for animals instead of what all too often it is.

From the beginning I never wanted the Black Beauty Ranch to stand alone. Just as we have always insisted that whatever animal we rescued had, as soon as possible, one of its own kind with him or her, so we felt about the Ranch itself. And all of us are now pleased that today there are many such sanctuaries as ours—indeed, now a whole association of them. Some of these handle certain kinds of animals, others, other kinds. Some are as small as backyard farms, others, almost as large as ours. But in all of them the animals, previously wild or not, live in that extraordinary gray area that lies between petdom and wilddom.

As I said earlier, Black Beauty Ranch is not a zoo. The animals we have rescued are not with us just to be looked at, but first and foremost to be looked after. At the beginning, when we were overwhelmed with the number of wild burros, wild goats, and wild pigs, we offered animals for adoption as long as the adopters met all our demands—that they not keep one animal alone, and that they not keep them for riding or as beasts of burden. And although we had many hundreds of these successful adoptions, we now do comparatively few. The adopters must really evince to us an extraordinary interest in the animals they want—people who have so fallen for one or more of them that they come over and over to see him or her and to be with them. It is not that we are selfish hoarders of our animals. It is, rather, that so many of our animals came to us, in the beginning, abused or ill-used that we do not want to take even the remotest chance that such misfortune would ever happen to any of them again.

Those of you who do come to visit the Ranch, and I am happy so many have, will find nothing that will harm the animals—no junk food, no rides, no shows. We do offer, however, a weekly open house, and for this we charge no admission. You can get into Black Beauty for nothing, and, unless you are unlucky enough to get me as a tour leader, you can even get out for nothing. With me, however (and just between us), I would not bet on it.

MORE FROM CLEVELAND AMORY

THE CAT WHO CAME FOR CHRISTMAS

The bestselling cat book of all time, *The Cat Who Came for Christmas* is the heartwarming and often hilarious story of a self-confessed curmudgeon and the stray cat he rescued on Christmas Eve. This classic tale of a proud, wary, but lovable cat and the cantankerous but funny and much-loved writer whom he owns has won millions of fans all year round, and especially during gift-giving season.

"You will smile all year . . . Cleveland Amory has written a book of delights." —Paul Harvey

ISBN 0-14-025273-8

FOR THE BEST IN PAPERBACKS, LOOK FOR THE 

In every corner of the world, on every subject under the sun, Penguin represents quality and variety—the very best in publishing today.

For complete information about books available from Penguin—including Puffins, Penguin Classics, and Arkana—and how to order them, write to us at the appropriate address below. Please note that for copyright reasons the selection of books varies from country to country.

In the United Kingdom: Please write to *Dept. JC, Penguin Books Ltd, FREEPOST, West Drayton, Middlesex UB7 0BR.*

If you have any difficulty in obtaining a title, please send your order with the correct money, plus ten percent for postage and packaging, to *P.O. Box No. 11, West Drayton, Middlesex UB7 0BR*

In the United States: Please write to *Consumer Sales, Penguin USA, P.O. Box 999, Dept. 17109, Bergenfield, New Jersey 07621-0120.* VISA and MasterCard holders call 1-800-253-6476 to order all Penguin titles

In Canada: Please write to *Penguin Books Canada Ltd, 10 Alcorn Avenue, Suite 300, Toronto, Ontario M4V 3B2*

In Australia: Please write to *Penguin Books Australia Ltd, P.O. Box 257, Ringwood, Victoria 3134*

In New Zealand: Please write to *Penguin Books (NZ) Ltd, Private Bag 102902, North Shore Mail Centre, Auckland 10*

In India: Please write to *Penguin Books India Pvt Ltd, 706 Eros Apartments, 56 Nehru Place, New Delhi 110 019*

In the Netherlands: Please write to *Penguin Books Netherlands bv, Postbus 3507, NL-1001 AH Amsterdam*

In Germany: Please write to *Penguin Books Deutschland GmbH, Metzlerstrasse 26, 60594 Frankfurt am Main*

In Spain: Please write to *Penguin Books S. A., Bravo Murillo 19, 1° B, 28015 Madrid*

In Italy: Please write to *Penguin Italia s.r.l., Via Felice Casati 20, I-20124 Milano*

In France: Please write to *Penguin France S. A., 17 rue Lejeune, F–31000 Toulouse*

In Japan: Please write to *Penguin Books Japan, Ishikiribashi Building, 2–5–4, Suido, Bunkyo-ku, Tokyo 112*

In Greece: Please write to *Penguin Hellas Ltd, Dimocritou 3, GR–106 71 Athens*

In South Africa: Please write to *Longman Penguin Southern Africa (Pty) Ltd, Private Bag X08, Bertsham 2013*